D0139229

Visualization
in
Teaching and Learning Mathematics

A project sponsored by the
Committee on Computers in Mathematics Education
of
The Mathematical Association of America

Editors

Walter Zimmermann
University of the Pacific

Steve Cunningham
California State University Stanislaus

MAA Notes Series

The MAA Notes Series, started in 1982, addresses a broad range of topics and themes of interest to all who are involved with undergraduate mathematics. The volumes in this series are readable, informative, and useful, and help the mathematical community keep up with developments of importance to mathematics.

Editorial Board
Warren Page, Chair

Donald W. Bushaw	Joan P. Hutchinson
Richard K. Guy	Vera S. Pless
Frederick Hoffman	David A. Smith

1. Problem Solving in the Mathematics Curriculum,
   Committee on the Undergraduate Teaching of Mathematics, *Alan Schoenfeld*, Editor.

2. Recommendations on the Mathematical Preparation of Teachers,
   CUPM Panel on Teacher Training.

3. Undergraduate Mathematics Education in the People's Republic of China,
   Lynn A. Steen, Editor.

4. Notes on Primality Testing and Factoring,
   by *Carl Pomerance*.

5. American Perspectives on the Fifth International Congress on Mathematics Education,
   Warren Page, Editor.

6. Toward a Lean and Lively Calculus,
   Ronald Douglas, Editor.

7. Undergraduate Programs in the Mathematical and Computer Sciences: 1985-86,
   D. J. Albers, R. D. Anderson, D. O. Loftsgaarden, Editors.

8. Calculus for a New Century,
   Lynn A. Steen, Editor.

9. Computers and Mathematics: The Use of Computers in Undergraduate Instruction,
   D. A. Smith, G. J. Porter, L. C. Leinbach, and R. H. Wenger, Editors.

10. Guidelines for the Continuing Mathematical Education of Teachers,
    Committee on the Mathematical Education of Teachers.

11. Keys to Improved Instruction by Teaching Assistants and Part-Time Instructors,
    Committee on Teaching Assistants and Part-Time Instructors, Bettye Anne Case, Editor.

12. The Use of Calculators in the Standardized Testing of Mathematics,
    John Kenelly, Editor.

13. Reshaping College Mathematics,
    Lynn A. Steen, Editor.

14. Mathematical Writing,
    by Donald E. Knuth, Tracy Larrabee, and Paul M. Roberts.

15. Discrete Mathematics in the First Two Years,
    Anthony Ralston, Editor.

16. Using Writing to Teach Mathematics,
    Andrew Sterrett, Editor.

17. Priming the Calculus Pump: Innovations and Resources,
    Thomas W. Tucker, Editor.

18. Models for Undergraduate Research in Mathematics,
    Lester Senechal, Editor.

First Printing
© 1991 by the Mathematical Association of America
Library of Congress number: 90-063690
ISBN 0-88385-071-0
Printed in the United States of America

Visualization
in
Teaching and Learning
Mathematics

Acknowledgements

This volume is sponsored by the Committee on Computers in Mathematics Education (CCIME) of the Mathematical Association of America. We would like to give special thanks to David A. Smith and Eugene Herman, chairs of the CCIME during the development of this volume. We also are grateful to Warren Page, MAA Notes Editor, and Don Albers, Chair of the MAA Publications Committee, for their advice and support.

This volume grew out of the editors' conviction that visual thinking and the development of visual tools through computer graphics could make major contributions to mathematics education. The development of the volume extended over two years from inception to publication. During this period, many persons helped to shape the volume through their thoughtful comments and suggestions. Among those whom we would like to thank for these contributions are many members of the CCIME, of the MAA Notes Editorial Board, the reviewers of the submitted papers, and the authors themselves.

The development of this volume was supported by a grant from the Education Committee of ACM-SIGGRAPH, the Special Interest Group on Computer Graphics of the Association for Computing Machinery, and we thank them for their generous support. We also thank our respective universities for their patience and support of our work.

The cover image is a Steiner surface produced by the **Vector** program described in the paper in this volume by Tom Banchoff and his students. We are grateful to Professor Banchoff for permission to use this image.

Walter Zimmermann
Steve Cunningham

October 10, 1990

Table of Contents

Editors' Introduction:
What is Mathematical Visualization?

Walter Zimmermann and Steve Cunningham

In the preface to *Geometry and the Imagination* [1], Hilbert wrote:

"In mathematics ... we find two tendencies present. On the one hand, the tendency toward abstraction seeks to crystallize the *logical* relations inherent in the maze of material that is being studied, and to correlate the material in a systematic and orderly manner. On the other hand, the tendency toward *intuitive understanding* fosters a more immediate grasp of the objects one studies, a live rapport with them, so to speak, which stresses the concrete meaning of their relations.

"...With the aid of visual imagination [*Anschauung*] we can illuminate the manifold facts and problems of geometry, and beyond this, it is possible in many cases to depict the geometric outline of the methods of investigation and proof ... In this manner, geometry being as many faceted as it is and being related to the most diverse branches of mathematics, we may even obtain a summarizing survey of mathematics as a whole, and a valid idea of the variety of its problems and the wealth of ideas it contains."

This passage effectively captures the spirit and purpose of this volume. Paraphrasing Hilbert, it is our goal to explore how "with the aid of visual imagination" one can "illuminate the manifold facts and problems of mathematics." We take the term *visualization* to describe the process of producing or using geometrical or graphical representations of mathematical concepts, principles or problems, whether hand drawn or computer generated. Our goal is to provide an overview of research, analysis, practical experience, and informed opinion about visualization and its role in teaching and learning mathematics, especially at the undergraduate level.

The Visualization Renaissance

The sciences, engineering, and to a more limited extent mathematics are enjoying a renaissance of interest in visualization. To a significant degree, this renaissance is being driven by technological developments. Computer graphics has greatly expanded the scope and power of visualization in every field. A recent report of the Board on Mathematical Sciences of the National Research Council [2] identifies a number of recent research accomplishments and related opportunities in mathematics. One of the areas cited is "Computer Visualization as a Mathematical Tool." According to this report, "In recent years computer graphics have played an increasingly important role in both core and applied mathematics, and the opportunities for utilization are enormous."

Walter Zimmermann is Professor and Chair of the Department of Mathematics at the University of the Pacific. He is an author of educational software for mathematics, and is a member of the Committee on Computers in Mathematics Education of the MAA.

Steve Cunningham is Professor of Computer Science at California State University Stanislaus. He has written widely on issues in computer graphics for education and is actively involved in both computer science and computer graphics education.

A report to the National Science Foundation, *Visualization in Scientific Computing* (ViSC) [3], calls for a national initiative "to get visualization tools into the 'hands and minds' of scientists." The report asserts that "Visualization ... transforms the symbolic into the geometric, enabling researchers to observe their simulations and computations. Visualization offers a method of seeing the unseen. It enriches the process of scientific discovery and fosters profound and unexpected insights. In many fields it is already revolutionizing the way scientists do science." The focus of the ViSC report is on the applied sciences, not mathematics. However, we would argue that much of the rationale of the report applies to mathematics as well as the sciences, and to teaching as well as research.

Visualization and the Nature of Mathematics

Hilbert refers to the two "tendencies" present in mathematics, logic and intuition. In their paper, "Nonanalytic Aspects of Mathematics and Their Implications for Research and Education," Philip J. Davis and James A. Anderson [4] observe: "Mathematics has elements that are spatial, kinesthetic, elements that are arithmetic or algebraic, elements that are verbal, programmatic. It has elements that are logical, didactic and elements that are intuitive, or even counter-intuitive... These may be compared to different modes of consciousness. To place undue emphasis on one element or group of elements upsets a balance. It results in an impoverishment of the science and represents an unfulfilled potential. ...We must not block off any mode of experience or thought." They go on to make a series of suggestions for implementing this philosophy. These recommendations include (among others): "Restore geometry. Restore intuitive and experimental mathematics. Give a proper place to computing and programmatics. Make full use of computer graphics." This volume is conceived very much in the spirit of these recommendations.

Visualization is not a recent invention. In "Picture Puzzling," Ivan Rival [5] writes, "Diagrams are, of course, as old as mathematics itself. Geometry has always relied heavily on pictures, and, for a time, other branches of mathematics did too. Even Isaac Newton ... did not actually prove [the] fundamental theorems. ...Had you asked him to justify them, he would likely have presented an argument that, though compelling, was loose and depended heavily on pictures."

Rival goes on to observe that "the intuitive and often persuasive style of argument used by Newton and his contemporaries fell into disrepute during the nineteenth century after it proved, in several celebrated cases, to be misleading." Davis and Anderson (in the paper quoted above) describe this process by the colorful phrase, "The Degradation of the Geometric Consciousness." They observe that "...over the past century and a half there has been a steady and progressive degradation of the geometric and kinesthetic elements of mathematical instruction and research. During this period the formal, the symbolic, the verbal, the analytic elements have prospered greatly."

The Science of Patterns

There is evidence that the pendulum has recently begun to swing back to a more balanced view of mathematics which takes into account more fully the visual and intuitive dimensions. In the article cited above, Rival states that "the limits of deductivism are at last dawning on mathematicians, thanks largely to computers." The role of computers in reshaping our notions of mathematics is echoed by Lynn Steen. In "The Science of Patterns" [6], Steen writes, "Mathematics is often defined as the science of space and number, as the discipline rooted in geometry and arithmetic. Although the diversity of modern mathematics has always exceeded this definition, it was not until the recent resonance of computers and mathematics that a more apt definition became fully apparent. Mathematics is the science of patterns. The mathematician seeks patterns in number, in space, in science, in computers, and in imagination. Mathematical theories explain the relations among patterns; functions and maps, operators and morphisms bind one type of pattern to another to yield lasting mathematical structures. Applications of mathematics use these patterns to 'explain' and predict natural phenomena that fit the patterns. Patterns suggest other patterns, often yielding patterns of patterns."

In speaking of "patterns," Steen is speaking metaphorically, but the metaphor of patterns is surely a visual metaphor. Not all patterns can be visualized, but it is as natural to want to visualize a pattern as it is to want to hear a melody. If mathematics is the science of patterns, it is natural to try to find the most effective ways to visualize these patterns and to learn to use visualization creatively as a tool for understanding. This is the essence of mathematical visualization.

Visualization and Understanding

The term "visualization" is unfamiliar to some in the context of mathematics, and its connotations may not be obvious. Our use of the term differs somewhat from common usage in everyday speech and in psychology, where the meaning of visualization is closer to its fundamental meaning, "to form a mental image." For example, there are psychological studies which focus on the subject's ability to form and manipulate mental images. In these studies, there is no question of using pencil and paper, much less a computer, to answer the questions. From the perspective of mathematical visualization, the constraint that images must be manipulated mentally, without the aid of pencil and paper, seems artificial. In fact, in mathematical visualization what we are interested in is precisely the student's ability to draw an appropriate diagram (with pencil and paper, or in some cases, with a computer) to represent a mathematical concept or problem and to use the diagram to achieve understanding, and as an aid in problem solving. In mathematics, visualization is not an end in itself but a means toward an end, which is understanding. Notice that, typically, one does not speak about visualizing a *diagram* but visualizing a *concept* or *problem*. To visualize a diagram means simply to form a mental image of the diagram, but to visualize a problem means to understand the problem in terms of a diagram or visual image. Mathematical visualization is the process of forming images (mentally, or with pencil and paper, or with the aid of technology) and using such images effectively for mathematical discovery and understanding.

Anschauung *and Intuition*

The title of Hilbert's book, quoted at the beginning of this article, is *Anschauliche Geometrie* [7]. The complexity (and ambiguity) of the term "visualization" reflects the richness of the related German noun *Anschauung*, from which the adjective *anschaulich* is derived. In *Imagery in Scientific Thought* [8], Miller points out that etymologically, *Anschauung* refers to visual perception, but Kant extended its use to cover any sort of perception, and subsequently the term took on a meaning often translated as *intuition*. In one context, Miller says, "...*Anschauung* refers to the intuition through pictures formed in the mind's eye." and "*Anschauung* is superior to viewing merely with the senses." (p. 110). In another context, *Anschauung* is "knowledge obtained by contemplation of ideas already in the mind..." (p. 274). Miller does not address mathematical visualization directly, but we would like to borrow his language to say that mathematical visualization also involves "intuition through pictures formed in the mind's eye." Thus *Anschauliche Geometrie,* literally rendered *visualizable geometry*, is also *intuitive geometry,* and by implication visualizable mathematics is also intuitive mathematics.

The ViSC report points out that "An estimated 50 percent of the brain's neurons are associated with vision. Visualization in scientific computing aims to put that neurological machinery to work." This interesting statement nevertheless blurs an important distinction because of the way it slips from *vision* to *visualization.* Miller's observation that "*Anschauung* is superior to viewing merely with the senses," implies that visualization is superior to vision. Visualization implies understanding — the kind of understanding that comes from "intuition through pictures formed in the mind's eye." Furthermore, in mathematics as well as in scientific computing, one may visualize something that is not seen or may never have been seen. Visualization may be "knowledge obtained by contemplation of ideas already in the mind... ." Many mathematicians can recall the experience of having an image come spontaneously to mind in the course of solving a problem — an image of some object or figure which they may have never actually seen.

Mathematical visualization is not "math appreciation through pictures." The intuition which mathematical visualization seeks is not a vague kind of intuition, a superficial substitute for understanding, but the kind of intuition which penetrates to the heart of an idea. It gives depth and meaning to understanding, serves as a reliable guide to problem solving, and inspires creative discoveries. To achieve this kind of understanding, visualization cannot be isolated from the rest of mathematics. Visual thinking and graphical representations must be linked to other modes of mathematical thinking and other forms of representation. One must learn how ideas can be represented symbolically, numerically, and graphically, and to move back and forth among these modes. One must develop the ability to choose the approach most appropriate for a particular problem, and to understand the limitations of these three dialects of the language of mathematics. We have thus strongly encouraged the authors of these papers to show how visualization operates in a mathematical context and not as an isolated topic.

Computing and Visualization

Steen and others have noted the role of computing in reshaping our concept of the nature of mathematics. Computers have a direct and concrete role in this visualization renaissance because of the ways computers can generate mathematical graphics. The beautiful and complex fractal images with which we are now familiar come to mind, but the images that come from mathematical visualization are varied indeed. They include geometrical figures of all kinds in two or three dimensions; they include curves and surfaces, direction fields, contour plots and other similar figures; they include graphs (in the sense of graph theory) and other kinds of schematic diagrams, such as Venn diagrams. The images being described need not be static; they may be dynamic or interactive (user controlled). Graphic simulations of processes, real or hypothetical, may belong to mathematical visualization, regardless of the nature of the images.

To encompass the whole scope of visualization, one should consider other technologies such as videotape, film, and interactive videodisc. These technologies further expand the kinds of images that can be used. Visualization technology is evolving so rapidly that it is impossible at this time to sketch its boundaries. The same is true about mathematical applications of visualization. Who could have predicted the explosion of interest in chaotic dynamics and the associated imagery of fractals? What will tomorrow bring?

There are many important questions about the role of computers in visualization. How can the power of computers in general, and interactive computer graphics in particular, be used most effectively to promote mathematical insight and understanding? What are the characteristics of good educational software? What are the roles of classroom demonstrations, structured computer laboratory exercises, and free exploration of mathematical ideas? What will be the impact of computers on the mathematics curriculum? In other words, how can computers help us to teach what we now teach more effectively, and what new problems, topics, or fields of mathematics are opened up by the new technology? Many of these questions are addressed, directly or indirectly, in the papers presented here.

Computer-based visualization, whether static, dynamic, or interactive, is only one facet of the role of computers in mathematics. Visualization must be linked to the numerical and symbolic aspects of mathematics to achieve the greatest results. Some of the most interesting and important applications of visualization involve problems which use numerical and/or symbolic processing as well as graphics.

A Paradigm for Using Color Graphics

Recent developments in the theory of nonlinear dynamical systems have not only dramatized the power of computer graphics but have established a new paradigm for mathematical visualization. Because of the crucial role of fractal images and other graphics in the study of dynamical systems, this subject has become a model, in general and specific terms, for other fields. For example, this model is directly acknowledged in a recent paper by Coffey, et al. [9], who write: "Several developments underlie the present revival in classical mechanics, including computerized algebraic processors and color graphics. ...Color graphics proves invaluable in visualizing the global behavior and discovering minutiae of local behavior hidden beneath the mass of calculation. Pseudocoloring a function over a domain, a widespread technique in applied mathematics, has produced stunning pictures; they have opened the eyes of mathematicians to hitherto unsuspected phenomena in the dynamics of nonlinear maps. Extension to classical mechanics forces a search for refinements in the technique such as automatic selection of colors to ensure enough contrast around isolated but close singularities."

Because of the importance of computers in visualization, it is natural to identify the field of visualization exclusively with computer-based visualization. In the context of mathematics, this would be a mistake. We construe the field of mathematical visualization broadly to include non computer-based visualization as well as visualization based on computers or other technologies. The ability to draw a simple figure to represent a mathematical problem, to interpret such figures with understanding, and to use such figures as an aid in problem solving are fundamental visualization skills. Without such fundamental skills, it is unlikely that computer-based visualization can be used efficiently, or even meaningfully. Vision is not visualization; to see is not necessarily to understand.

The Origins and Objectives of this Volume

The literature of mathematics includes works which address the graphical representation of mathematical ideas. One noteworthy recent example is George Francis' *A Topological Picturebook* [10]. Significantly, Francis' inspiration did not come from modern computer graphics but from the German *Enzyclopaedie der Mathematischen Wissenschaften* and the collected works of Felix Klein. He writes, "Theirs was a wonderfully straightforward way of looking at rather complicated things, notably Riemann surfaces and geometrical constructions over the complex numbers. They drew pictures, built models and wrote manuals on how to do this. And so they also captured a vivid record of the mathematics of their day."

The tradition of the *Enzyclopaedie* has been nearly lost, and works such as Francis' are rare. When visualization is discussed today, it is usually in the context of another subject. Sometimes, graphics is treated as one of the features of so-called "computer algebra systems" (CAS) and discussions of visualization are subsumed under the rubric of "symbolic computation." We consider this practice unfortunate. Symbolic computation, numerical computation and graphics are three complementary modes of mathematical computing, and to refer to computer programs which have all these capabilities as "computer algebra systems" is misleading. Terminology aside, we believe that visualization involves an interesting and distinctive set of mathematical, pedagogical and practical questions which ought to be addressed directly, not just as a side-issue in another context. This belief led us to undertake the publication of this volume.

The Organization of this Volume

This volume explores the role of visualization in mathematics education, especially undergraduate education. The first five papers address fundamental visualization issues which cut across different subject fields. The remaining papers address the role of visualization in particular fields of mathematics — geometry, calculus, differential equations, differential geometry, linear algebra, numerical analysis, complex analysis, stochastic processes and statistics. Some of these papers also address general visualization issues as they arise in the context of their fields. With two exceptions (the papers by Goldenberg), the papers are original papers prepared for this volume. These two papers are included at the request of the editors (and, of course, with permission), because of the contribution we feel they make.

There are few clear precedents or contemporary standards for publications in the field of mathematical visualization. We exercised our best judgment in selecting and refining the papers which appear here, and we accept responsibility for the tone and scope of the volume. While these papers have mathematical content, they are not mathematics papers in the traditional sense, nor are they research papers in education. The tone of the papers, on the whole, is informal, and we consider this to be appropriate for the subject matter and to be consistent with the tradition of the *MAA Notes*. Unlike formal scholarly papers, some of these papers include opinion and speculation. This, too, we consider appropriate, as the field is rapidly evolving.

An issue we had to face was how to deal with papers which discuss software, especially software developed by the authors. It is natural for someone who has interests in visualization to be drawn to software development, and those who have developed and implemented software are among those best qualified to discuss visualization. We considered it inappropriate to include papers whose goal seemed to be promotion of particular software. However, we felt it would be counterproductive to exclude papers just because they discuss the author's software. We made a conscious decision to include such papers where the focus of the paper is on mathematical or pedagogical issues, not on the software, and where any discussion of the software is in the context of such issues. Several papers in this volume include such discussion, and we are satisfied that they meet these criteria.

Broadly speaking, our goals in this project were to present ideas and information about the current state of mathematical visualization, to stimulate thought and discussion, and to provide a frame of reference for future efforts. We hope that this volume will contribute to a better understanding of the educational aspects of the field we have called mathematical visualization, and a broader recognition of its importance. We believe the papers should be judged by the extent to which they show how one can use visual imagination and visual thinking to "illuminate the manifold facts and problems" of mathematics. Do these papers, and the visual images they present, convey the "patterns" of mathematics? To what extent do these papers show how one can promote the development of intuition "through pictures formed in the mind's eye"?

Visualization, like geometry, is "related to the most diverse branches of mathematics." Visualization is multifaceted: rooted in mathematics, the field has important historical, philosophical, psychological, pedagogical and technological aspects. If present trends are any indication, it seems that mathematics will evolve in a direction which will make visualization even more important in the future than it is now. At the same time, the evolution of technology will make more and more powerful visualization tools available. We hope all readers will find ideas here which they can use in their own work and in their classes, and we hope some readers will be encouraged to pursue studies in this fascinating subject. The opportunities for creative efforts by mathematicians and others appear to be unlimited.

References

[1] Hilbert, David, and S. Cohn-Vossen, *Geometry and the Imagination*, Translated by P. Nemenyi. Chelsea, N.Y., 1983. *Note*: Although this is authored jointly by Hilbert and Cohn-Vossen, the Preface, from which the quoted passage was taken, is signed by Hilbert alone.

[2] National Research Council, *Renewing U.S. Mathematics*, National Academy Press, Washington, D.C., 1990.

[3] McCormick, Bruce H., Thomas A.DeFanti, and Maxine D. Brown (eds), Visualization in Scientific Computing, *Computer Graphics* 21(November 1987).

[4] Davis, Philip J. and James A. Anderson, "Nonanalytic Aspects of Mathematics and Their Implication for Research and Education," *SIAM Review* 21(1979), 112-117.

[5] Rival, Ivan, "Picture Puzzling: Mathematicians are Rediscovering the Power of Pictorial Reasoning," *The Sciences* 27(1987), 41-46.

[6] Steen, Lynn A., "The Science of Patterns," *Science* 29, April 1988.

[7] Hilbert, David and S. Cohn-Vossen, *Anschauliche Geometrie*, Springer, Berlin, 1932.

[8] Miller, Arthur I., *Imagery in Scientific Thought* , MIT Press, 1987.

[9] Coffey, Shannon, André Deprit, Étienne Deprit, and Liam Healy, "Painting the Phase Space Portrait of an Integrable Dynamical System," *Science* 247(1990), 833-836.

[10] Francis, George K., *A Topological Picturebook*, Springer, New York, 1987.

Acknowledgement

The work leading to this article was supported in part by the National Science Foundation under grant number DMS 8851255.

Visual Information and Valid Reasoning

Jon Barwise and John Etchemendy

Psychologists have long been interested in the relationship between visualization and the mechanisms of human reasoning. Mathematicians have been aware of the value of diagrams and other visual tools both for teaching and as heuristics for mathematical discovery. As the papers in this volume show, such tools are gaining even greater value, thanks in large part to the graphical potential of modern computers. But despite the obvious importance of visual images in human cognitive activities, visual representation remains a second-class citizen in both the theory and practice of mathematics. In particular, we are all taught to look askance at proofs that make crucial use of diagrams, graphs, or other nonlinguistic forms of representation, and we pass on this disdain to our students.

In this paper, we claim that visual forms of representation can be important, not just as heuristic and pedagogic tools, but as legitimate elements of mathematical proofs. As logicians, we recognize that this is a heretical claim, running counter to centuries of logical and mathematical tradition. This tradition finds its roots in the use of diagrams in geometry. The modern attitude is that diagrams are at best a heuristic in aid of finding a real, formal proof of a theorem of geometry, and at worst a breeding ground for fallacious inferences. For example, in a recent article, the logician Neil Tennant endorses this standard view:

> [The diagram] is only an heuristic to prompt certain trains of inference; … it is dispensable as a proof-theoretic device; indeed, … it has no proper place in the proof as such. For the proof is a syntactic object consisting only of sentences arranged in a finite and inspectable array.

It is this dogma that we want to challenge.

We are by no means the first to question, directly or indirectly, the logocentricity of mathematics and logic. The mathematicians Euler and Venn are well known for their development of diagrammatic tools for solving mathematical problems, and the logician C. S. Peirce developed an extensive diagrammatic calculus, which he intended as a general reasoning tool. Our own challenge is two-pronged. On the one hand, we are developing a computer program, *Hyperproof*, that follows in the tradition of Euler, Venn, and Peirce. The program will allow students to solve deductive reasoning tasks using an integrated combination of sentences and diagrams. On the other hand, we are developing an information-based theory of deduction rich enough to assess the validity of heterogeneous proofs, proofs that use multiple forms of representation. In this task, we do not want to restrict our attention to any particular diagrammatic calculus; rather, our aim is to develop a semantic analysis of valid inference that is not inextricably tied to linguistic forms of representation.

Jon Barwise is College Professor of Philosophy, Mathematics, and Logic at Indiana University. John Etchemendy is Director of the Center for the Study of Language and Information and Associate Professor of Philosophy and Symbolic Systems at Stanford University. Professors Barwise and Etchemendy have jointly authored The Liar: An essay on truth and circularity, have developed the courseware packages Turing's World and Tarski's World, and are leading the development of the Hyperproof project.

This volume is not an appropriate forum for presenting the technical details of our theory of heterogeneous inference. Nor can we actually illustrate our program in operation. Consequently, our aim in this paper is just to sow a seed of doubt in the reader's mind about the dogma mentioned above. In the next section we give a number of examples of heterogeneous inference, showing the wide range of types of reasoning that we think fall under this heading. In the third section, we give an example of the way *Hyperproof* can be used to solve a simple problem. In the concluding section, we present some thoughts about what it is that makes diagrams and other forms of visual representation so useful for mathematical discovery, proof, and pedagogy. In a companion paper (Barwise and Etchemendy [1]) we give a preliminary account of our information-based theory of deduction, and apply it to one of the examples from this paper.

The Legitimacy of Heterogeneous Inference

Valid deductive inference is often described as the extraction or explication of information that is only implicit in information already obtained. Modern logic builds on this intuition by modeling inference as a relation between sentences, usually sentences of a formal language like the first-order predicate calculus. In particular, it views valid deductive proofs as structures built out of such sentences by means of certain predetermined formal rules. But of course language is just one of the many forms in which information can be couched. Visual images, whether in the form of geometrical diagrams, maps, graphs, or visual scenes of real-world situations, are other forms.

A good way to appreciate the importance of nonlinguistic representation in inference is simply to look at some everyday examples of valid deductive reasoning. This exercise serves two purposes. First, it convinces us of the ubiquity of nonlinguistic representation in actual reasoning. Second, it calls into question some of the basic assumptions that have led to the theorist's disdain for such reasoning.

On the Universality of Linguistic Representation

The modern account of inference given to us by logic has an unstated assumption, namely, that all valid reasoning is (or can be) cast in the form of a sequence of sentences in some language. This picture has been strongly challenged by psychologists investigating visualization. An example tracing back to Stenning [4] and Kosslyn [2] involves the use of maps and map-like representations in human problem solving.

Example 1. Suppose you are a tourist in San Francisco's Chinatown, and a motorist stops and asks how to get to China Basin. You take out your map, find both Chinatown and China Basin, and tell him what route he could take.

Here there is a clear sense in which you have engaged in a valid piece of deductive reasoning, one whose assumptions consist in part of the information provided by the map and whose conclusion consists of the claim that a certain route will take the motorist to China Basin. But notice that you do not simply "read" the map in the sense in which you might read the same directions written out on a scrap of paper. The map contains a vast amount of information, most of which is useless for any particular use of the map. In particular, it contains information about how to get from Chinatown to China Basin, the information also contained in your subsequent directions. But this information has been extracted from the map and expressed in a very different form.

Kosslyn uses such examples to argue that visual and linguistic representations of the same information have different properties. His point is that one form may represent a given piece of information more efficiently than another for a particular purpose. This same conclusion is reinforced by Stigler [5] in studying the role of visualization in mental calculation. Noting Kosslyn, Stigler observes that "much of what we know can be represented in more than one format (e.g., images or propositions), and … for different tasks, different representational formats may lead to differences in the nature, speed, and efficiency of the processing they support" (p. 175).

We think this is an important point. But we also think that in many cases it is more than a matter of alternative representations of a given piece of information. Consider another example.

Example 2. Suppose you are at a large party and want to meet a visiting mathematician (Anna, to give her a name) whom you know to be there. You are told that Anna is in the next room talking to a man with a beard. When you enter the room you see that, unfortunately, there are two bearded men. Fortunately, though, they are both talking to the same woman. You conclude that she must be Anna.

Note that in this case your conclusion is based on two forms of information. One of these is the earlier statement (plus your knowledge that Anna is a woman), the other is information you get from the scene in front of you. Although some might claim that you translate the latter information into mental sentences of some sort, there is no obvious reason to think this is so.

For our purposes, the crucial feature of this example is that the conclusion associates a name with a person in a way that transcends each domain individually, both the linguistic and the visual. Because of this, the reasoning cannot be accurately modeled by deductions in a standard formal language. The nearest sentential analog to this conclusion might associate a name with some description ("Anna is the woman who …"), rather than with Anna herself. Alternatively, we might employ some deictic, demonstrative, or indexical element ("That woman is Anna"). But of course it is not this sentence, in isolation, that is the conclusion of your reasoning. Only when we interpret the demonstrative as referring to Anna have we captured the genuine content of your conclusion.

We also note that if this really is a case of heterogeneous inference, then it is clear that this kind of reasoning is ubiquitous in daily life.

On the Dangers of Visual Representation

The main reason for the low repute of diagrams and other forms of visual representation in logic is the awareness of a variety of ill-understood mistakes one can make using them: witness the fallacies that have arisen from the misuse of diagrams in geometry. By contrast, it is felt that we have a fairly sophisticated semantic analysis of linguistically based reasoning.

Our counter to this attitude is two-fold. First, we note the obvious fact that a wide variety of mistaken proofs and fallacious inferences do not use visual information. These range from the traditional informal fallacies, to the misapplication of formal rules (for example, inadvertently capturing a free variable in a substitution), to mistakes far harder to classify or categorize. And, as we all know, a simple diagram can often be used to pinpoint such an error. The mere existence of fallacious proofs is no more a demonstration of the illegitimacy of diagrams in reasoning than it is of the illegitimacy of sentences in reasoning. Indeed, what understanding we have of illegitimate forms of linguistic reasoning has come from careful attention to this form of reasoning, not because it was self-evident without such attention.

Our second reply is to point out that although one can make mistakes using various forms of visual representation, it is also possible to give perfectly valid proofs using them. We give some examples here. In the companion paper [1] we present an information-based analysis of inference and illustrate its use by analyzing the validity of Example 5 below.

Example 3. Recall that the Pythagorean Theorem claims that given any right triangle, the sum $a^2 + b^2$ is equal to c^2, the square of the hypotenuse. One familiar proof of this theorem involves a construction that first draws a square on the hypotenuse, and then replicates the original triangle three times as shown. Using the fact that the sum of the angles of a triangle is a straight line, one easily sees that $ABCD$ is itself a square, one whose area can be computed in two different ways. On the one hand, its area is $(a + b)^2$, since the sides of the large square have length $a + b$. On the other hand, we see by inspection that its area is also $c^2 + 4x$, where x is the area of the original triangle, i.e., $x = ab/2$. This gives us the equation

$$(a + b)^2 = c^2 + 2ab,$$

which leads to the desired equation,

$$a^2 + b^2 = c^2.$$

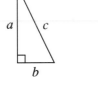

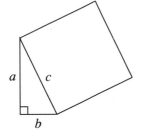

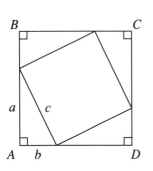

It seems clear that this is a legitimate proof of the Pythagorean theorem. Note, however, that the diagrams play a crucial role in the proof. We are not saying that one could not give an analogous (and longer) proof without them, but rather that the proof as given makes crucial use of them. To see this, we only need note that without them, the proof given above makes no sense.

This proof of the Pythagorean theorem is an interesting combination of both geometric manipulation of a diagram and algebraic manipulation of nondiagrammatic symbols. Once you remember the diagram, however, the algebraic half of the proof is almost transparent. This is a general feature of many geometric proofs: Once you have been given the relevant diagram, the rest of the proof is not difficult to figure out. It seems odd to forswear nonlinguistic representation and so be forced to mutilate this elegant proof by constructing an analogous linguistic proof, one no one would ever discover or remember without the use of diagrams.

As we have noted, there is a well-known danger associated with the use of diagrams in geometrical proofs. The danger stems from the possibility of appealing to accidental features of the specific diagram in the proof. For example, if any piece of our reasoning had appealed to the observation that, in our diagram, a is greater than b, the proof would have been fallacious, or at any rate it would not have been as general as the theorem demands. Nevertheless, it is clear that we did not make use of any such accidental features in our proof. Further, it should be noted that a linguistically presented proof can have accidental features that lead to errors as well. For example, one of the constituent sentences may be ambiguous. More than one error has resulted from a valid step that is expressed ambiguously.

The potential for error in diagrammatic reasoning is real. But as we have noted, it is no more serious than the sorts of fallacies that can occur in purely linguistic forms of reasoning. The tradition has been to address these latter fallacies by delving into the source of the problem, developing a sophisticated understanding of linguistic proofs. It is not obvious that an analogous study of diagrammatic reasoning could not lead to an analogous understanding of legitimate and illegitimate uses of these techniques.

Example 4. We are all familiar with the use of Venn diagrams to solve problems and illustrate theorems in set theory. Consider, for example, their use in proving the distributive law,

$$A \cup (B \cap C) = (A \cup B) \cap (A \cup C)$$

To prove this, we draw two copies of a diagram with circles used to represent the three sets. We successively shade in these diagrams, with operations corresponding to union and intersection, first for the left-hand side of the equation, then for the right-hand side. The fact that we end up with the same diagram in both cases shows that the distributive law holds.

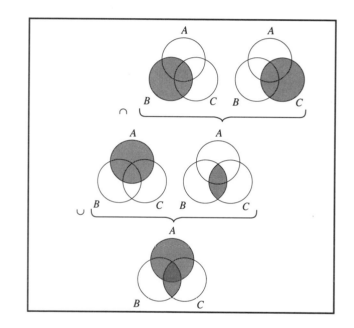

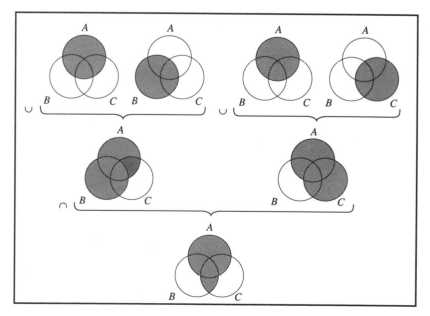

Venn diagrams provide us with a formalism that consists of a standardized system of representations, together with rules for manipulating them. In this regard, they could be considered a primitive visual analog of the formal systems of deduction developed in logic. Indeed, it is possible to give an information-theoretic analysis of this system, one that shows, among other things, that the demonstration above is in fact a valid proof.

In Venn diagrams, regions on the page represent sets, adding and intersecting regions correspond to union and intersection of sets. Shading a region serves to focus attention on that region. What makes our purported proof a real proof is the fact that there is a homomorphic relationship between regions (with the operations of addition and intersection of regions) and sets (with the operations of union and intersection of sets).[1]

As we have said, diagrams can lead us into error if they are used improperly. For example, if we were to take the nonemptiness of a region as indicating nonemptiness of the represented set, we would be led to conclude that the intersection of A, B, and C is nonempty, which it might not be. Likewise, the size of a region carries no representational significance whatsoever about the represented set, any more than the size of the letter "A" indicates that the set denoted is larger than the set denoted by "a." The use of diagrams, like the use of linguistic symbols, requires us to be sensitive to the representational scheme at work.

Mathematics instructors know from experience that these sorts of diagrammatic proofs are much more helpful and convincing to students than more standard linguistic proofs. Frequently, they enable the student to see, in a way that formal proofs do not, just what the theorem is saying and why it is true.

The sort of reasoning we have examined most closely is typified by problems found in puzzle magazines or on the analytical reasoning section of the Graduate Record Examination. A very simple example of this kind of problem is described in Example 5.

Example 5. You are to seat four people, A, B, C, and D in a row of five chairs. A and C are to flank the empty chair. C must be closer to the center than D, who is to sit next to B. From this information, show that the empty chair is not in the middle or on either end. Can you tell who must be seated in the center? Can you tell who is to be seated on the two ends?

We urge the reader to solve this problem before reading on. As simple as it is, you will no doubt find that the reasoning has a large visual component. Probably you will find it useful to draw some diagrams. With more complex problems of this sort, diagrams become even more essential.

One line of reasoning that can be used in solving this problem (the one we analyze in [1]) runs as follows. Let us use the diagram below to represent the five chairs.

Our first piece of information tells us that *A* and *C* are to flank the empty chair. Let us use X to signify that a chair is empty. Then we can split into six cases. Or, since the problem does not distinguish left from right in any way, we can limit our attention to three cases, the other three being mirror images of them.

Case 1

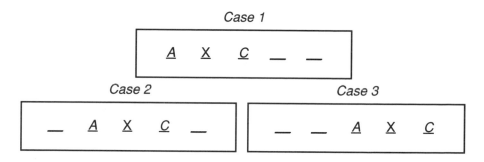

Case 2   *Case 3*

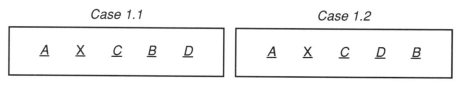

Using the fact that *C* must be seated closer to the center than *D*, we can eliminate Case 3, since *C* is not closer to the center than any available chair. Similarly, since *D* must sit next to *B*, we can rule out Case 2, since no contiguous chairs are available. This leaves us with the following two possibilities:

Case 1.1   *Case 1.2*

In both of these cases, all of the stated constraints are satisfied, and so we know that neither case can be ruled out. This allows us to answer the questions posed in the puzzle. First, we see that in both cases the empty chair is not in the middle or on either end, as desired. Second, we see that *C* must be seated in the middle. And finally, we see that *A* must be on one end of the row, but we do not know whether *B* or *D* is on the other end. Either one is possible.

We are convinced that, properly understood, the demonstration above is a valid proof of the conclusions reached, and not just a psychological crutch to help us find such a proof, as the traditional account would have us believe. Moreover, we think that there are some parts of this proof that are missed in traditional accounts of inference. Notice, for example, that although our reasoning is all of a piece, the three parts of the problem have quite different characteristics when looked at from the traditional perspective. The first question asks us to prove that a specific fact follows from the given information. The second, in contrast, asks us whether a certain sort of information is implicit in the given information, and the answer is *yes.* The third question is of the same sort, but here we end up showing that something does *not* follow from the given, namely, who is seated on the end opposite *A*. We show this *nonconsequence* result by coming up with models of the given, one of which has *B* on the end, while the other has *D*. Normally, showing that something follows from some assumptions and showing that something does not follow are thought of as dual tasks, the first deduction and the second model building. This dichotomy seems unhelpful in analyzing the reasoning above, since there was no apparent discontinuity between the portions of the reasoning that demonstrated consequence and the portions that demonstrated nonconsequence. This occurence, in one reasoning task, of parts that are usually thought of as duals of one another (deduction and model construction) turns out to be typical of a whole spectrum of reasoning tasks. The desire to account for this dual nature of ordinary reasoning is an important motivation for some of the details of our theory of heterogeneous inference.

We have claimed that diagrams and other forms of visual representation can be essential and legitimate components in valid deductive reasoning. As we have noted, this is a heretical claim, running counter to a long tradition in mathematics, logic, and certain traditions within cognitive psychology. No doubt many readers will initially find it implausible. We invite such readers to work out a solution to Example 5 using a standard system of deduction, and to compare the complexity and structure of that solution with the original solution using diagrams. This sort of exercise has convinced us that even when syntactic analogs can be constructed, they are poor models of much actual valid reasoning.

Various Uses of Visual Information in Valid Reasoning

Looking back at these and similar examples, we see that there are at least three different ways that visual representations can be part of valid reasoning. (i) Probably the most obvious and pervasive form in everyday life is that visual information is part of the given information from which we reason. In the simplest case, this sort of reasoning extracts information from a visual scene and represents it linguistically. (ii) Visual information can also be integral to the reasoning itself. At its most explicit, such reasoning will employ an actual diagram, as in the examples above. In other cases, there may be no need for such an explicit diagram, since the reader will be able to visualize the steps without it. (iii) Finally, visual representations can play a role in the conclusion of a piece of reasoning. Imagine replacing the story of Anna above with a similar story where the problem is to provide a caption for a photo identifying the people represented by the photo. In a given problem, visual information could play any or all of the above roles.

We suspect that visualization of sort (ii) plays a much bigger role in mathematical proofs than is generally acknowledged, and that this is in fact part of what accounts for the discrepancy between mathematical proofs and their formal counterparts. In calculus, for example, the concept of a continuous function is essentially a visual one, one that is linguistically captured by the usual ε-δ characterization. But it seems clear that in giving proofs we often shortcut this characterization and rely more directly on the visual concept. Another example is the ubiquity of arrow diagrams in modern mathematics. As Saunders Mac Lane puts it in the first sentence of his classic book on category theory:

> Category theory starts with the observation that many properties of mathematical systems can be unified *and simplified* by a presentation with diagrams of arrows.[2] (Mac Lane [3], p. 1)

If visualization plays a bigger role in mathematical proof than has been widely recognized by our accounts of proof, it would explain in part the difficulty of creating automated proof checkers, let alone automated proof generators, based as they are on the linguistic model of reasoning. And it could account for the trouble our students have in mastering the ability to prove things.

It might be thought that visual representations would only be appropriate in proofs of results that can themselves be visualized or visually represented in a natural way. However, this is not the case. For example, it has been suggested that what is paradoxical about the Banach-Tarski paradox rests in the fact that it is a result that defies our visual intuitions. This may well be right. However, it does not follow that one cannot use diagrams or other forms of visual representation effectively in giving its proof. Indeed, the most popular proofs of the result do use diagrams.

A striking feature of diagrammatic reasoning is its dynamic character. The reasoning often takes the form of successively adding to or otherwise modifying a diagram. This makes it a very convenient form for use in one-on-one discussions at a blackboard or in front of a class. However, it makes it very awkward to use in a traditional linguistically based document, say a book or research paper. Adding to this problem are the difficulties in getting even a static diagram into print at all. For example, each diagram in this paper took us much more time and effort than a similar amount of text would have taken. However, this need not be so, and indeed it is changing all the time. Computer technology is making it ever easier to create and replicate diagrams and other forms of visual information in an accurate and relatively painless manner. As this process continues, it will become increasingly convenient for mathematicians to use static diagrams in papers and texts.

Of course it is also possible to think beyond the static printed documents of the past, to dynamic computer-driven documents of the future. In these documents, it is possible to create dynamic visual representations that unfold before the reader's eyes. It seems likely that such documents will lead to proofs that would otherwise be impossible to find or comprehend. In the next section, we turn to a discussion of *Hyperproof*, a tiny first step along that road. Unfortunately, we are confined to the printed page, but we will try to give a feel for the dynamic program.

Hyperproof

Hyperproof allows the user to reason based on two forms of information about a blocks world. At the top of the screen, there is a diagram of the blocks world. These worlds consist of objects of various shapes and sizes located on an 8 by 8 grid. The possible shapes are cube, tetrahedron, and dodecahedron; the possible sizes are small, medium, and large. A crucial feature of the visual representation is that the depicted information can be incomplete in a variety of ways. For one thing, the objects depicted in the world may be named, but their names may not be indicated in the diagram. For another, an object's location, size, or shape may not be determined.[3]

Below the diagram, the user is given information by means of sentences in the usual first-order logical notation. This information will typically be compatible with, but go beyond, the incomplete information depicted in the diagram. It will also typically be the sort of information not easily incorporated in a diagram.

Given these two sources of information about a blocks world, the student is presented with some reasoning task or tasks. These come in a variety of forms, analogous to some of the examples in the preceeding section. In one type of problem, the student is asked to use the given information to identify one of the depicted objects by name. In another, the student is asked to determine whether some other claim follows from the given information. If it does, the student is required to demonstrate this by means of a proof, one that can use both sentences and diagrams that extend the original diagram. If it does not follow, the student is required to prove this by constructing an extension of the orginal diagram that depicts a world that falsifies the claim. More sophisticated reasoning tasks that combine these can also be given, such as "What is the most you can say about the number of cubes to the left of *d* ?" Such a problem might involve identifying *d*, showing that in all worlds compatible with the given information, there are at least three and at most four cubes to the left of *d*, and, finally, showing that in some such worlds there are exactly three, while in others there are exactly four.

Hyperproof provides, in addition to the facilities for presenting the sorts of information described above, a system of inference rules that allows the student to manipulate the information. The rules include all the usual linguistic rules familiar from first-order logic, but also rules for dealing with the extralinguistic information provided by the diagram. A "keyboard" for use in giving proofs can be found on the right of the screen. Rather than go into detail about these rules, we will illustrate them by means of an example.

Figure 1: *The initial screen*

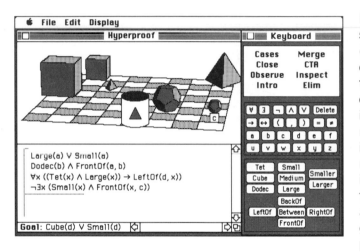

Figure 1 shows a sample problem that a student might solve using *Hyperproof*. The goal of this problem (stated at the lower left) is to prove that *d* is either a cube or small. The given information comes in the form of four sentences plus the information in the diagram. The diagram indicates that one of the objects is named *c* but does not indicate which objects are named *b* or *d*. And, as it happens, there is not enough information given to determine which objects are named *b* or *d*. Another way in which this diagram is incomplete has to do with the size of one of the tetrahedra. The "barrel" with a triangle on it represents a tetrahedron whose size is unknown. (Similar graphical devices are used to represent other forms of partial information.)

The reader might like to solve this problem before reading on. To do so, one needs to know that *FrontOf* means further forward though not necessarily on the same file, and *LeftOf* means to the viewer's left, though not necessarily on the same rank. We will work through one possible solution to this problem.

We start by citing the second sentence to break into two cases (using the rule **Exhaustive Cases**), as shown in Figure 2. In the first of these cases, we have labelled the medium-sized dodecahedron *b* and the tetrahedron in front of it *a*. In the second, we have labelled the other dodecahedron *b* and the tetrahedron in front of it *a*. The system permits this breaking into cases only

Figure 2: *After breaking into cases*

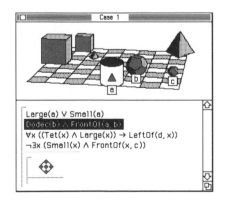
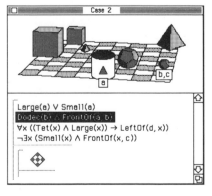

Figure 3: *After merging cases*

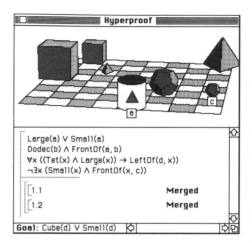

because every blocks world compatible with the original diagram and the cited sentence must also be compatible with one of these two extensions of the diagram. So, for example, had there been another possible dodecahedron in the diagram, these cases would not have been exhaustive, and the program would have required the student to find the missing case or cases.

At this stage, we are in a situation analogous to the Anna example in the preceding section. Namely, we are not sure which object is *b* but, in either case, we know that the tetrahedron of unknown size is *a*. This allows us to use the rule **Merge Cases** to regain a single diagram containing the information that *a* is the tetrahedron of unknown size. This is shown in Figure 3.

We are now in a position to use the information provided by the first sentence, that *a* is large or small, since we now know which object is *a*. We again use the rule **Exhaustive Cases** to break into cases, one where *a* is large, the other where *a* is small. The result is shown in Figure 4.

The second of these cases can be eliminated on the basis of the information contained in the last sentence. This sentence says that there

Figure 4: *After breaking into cases again*

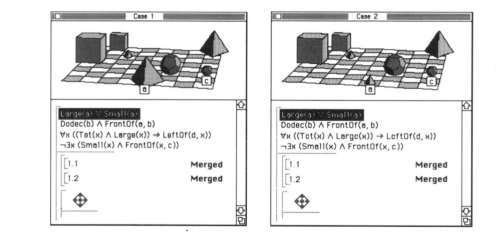

Figure 5: *Closing case 2*

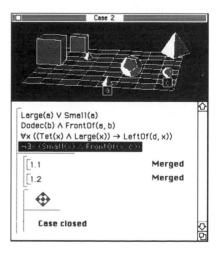

is nothing small in front of *c*, but the second diagram in Figure 4 indicates that *a* is both small and in front of *c*. We can thus apply the rule **Close Case** by citing the diagram and the fourth sentence, as shown in Figure 5. The program checks that the sentence is false in any blocks world compatible with the diagram, and so permits this closing, leaving us in the situation depicted in Figure 6.

We next cite the third sentence which says that *d* is to the left of every large tetrahedron. This allows us to use the rule **Exhaustive Cases** again to break into the three cases shown in Figure 7. Each of these cases is compatible with all the given information, so we cannot determine which object is *d*. But we observe that the desired conclusion, that *d* is either a cube or small, holds in each of these cases. Thus, we can use the rule **Inspect**, which allows us to survey all the open cases, verifying that a particular claim holds in all of them. The program displays each open case in turn, evaluates the desired claim, and indicates whether it is true or not before going on to the next case. If the

Figure 6: After closing case 2

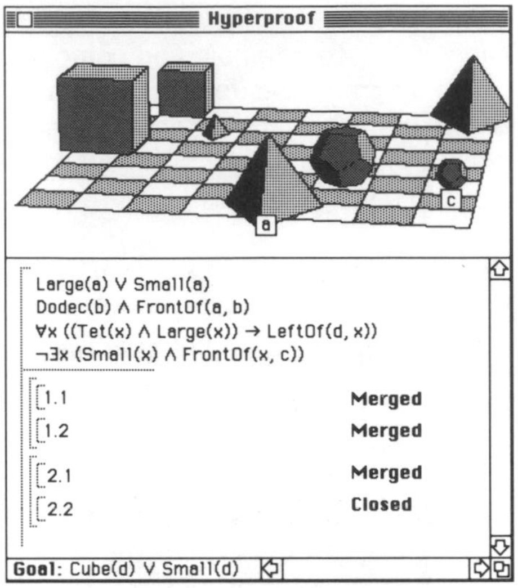

Figure 7: *After breaking into cases yet again*

Figure 8: *The completed proof*

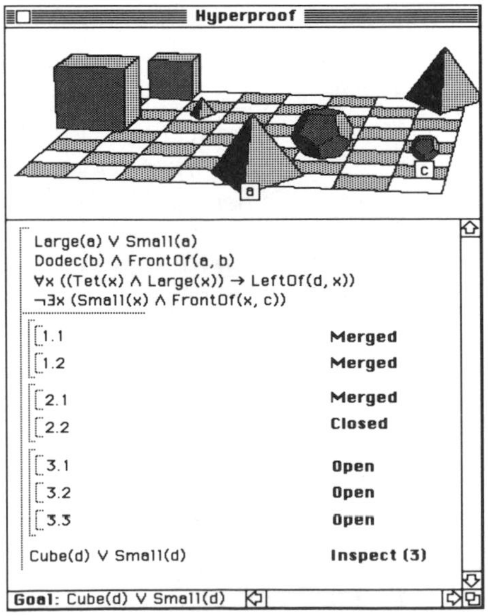

claim holds in all cases, the program indicates that the rule has been successfully applied. This leaves us with the final stage of our proof, as shown in Figure 8.

This example does not illustrate the use of several important rules available to the student. For example, it does not show any of the traditional logical rules of inference at work. But it does give a rough feel for the sorts of things one can do using the diagrams.

We have never had the patience to work out a purely linguistic proof of this problem. We have done this exercise for a much simpler problem, where the resulting proof was over 100 lines long, and was quite nonintuitive. This was a problem where one could think through the diagrammatic solution in one's head with little effort. The present problem is more difficult. We suspect that a purely linguistic proof would be on the order of 200 to 300 lines long, well beyond the reach of any but the most diligent logic student with no date for the weekend.

We believe that a program like *Hyperproof* will aid greatly in teaching students general reasoning skills of use both in mathematics and in everyday life. In part it works by forcing the student to focus on the subject matter of the problem, rather than to resort to purely formal manipulation of symbols. But it also works by exploiting the power of visual representations in reasoning.

The system we are currently developing is restricted to reasoning about blocks worlds. However, a second generation program is in the planning stage. It will provide a toolbox for constructing diagrams of a more abstract and general sort, for use in solving a wide variety of puzzles and problems, typified by the analytic reasoning problems found on the Graduate Record Examination. In particular, it would allow us to carry out the solution given to Example 5.

Inference as Information Extraction

During the past hundred years, logicians have developed an arsenal of techniques for studying valid inference in the case where all the information is expressed in sentences of a formal language. On the one hand, there are many varieties of deductive systems, ranging from natural deduction to Gentzen's sequent calculus to Hilbert-style formalisms to semantic tableaux. On the other hand, semantic techniques have been developed that allow us to assess the adequacy of given deductive systems. The most widely accepted semantic analysis of consequence represents the information content of a sentence with the collection of structures in which the sentence is true, and declares one sentence to be a consequence of another if the information content of the first is contained in the content of the second, that is, if all the models of the second are models of the first. A deductive system is said to be sound provided that a derivation of one sentence from another is possible only if it is in fact a consequence of that sentence. The system is said to be complete if whenever one sentence is a consequence of another there is in fact a derivation of the one from the other.

These familiar techniques have been tailor made to the homogeneous, linguistic case. When we turn our attention to heterogeneous inference, we need to generalize them along various dimensions. On the one hand, it would be nice to exhibit something comparable to a deductive system that allows us to generate mathematical analogs of valid reasoning, but where the reasoning involves multiple forms of representation. But more importantly, we need a semantic touchstone to assess the adequacy of such reasoning.

In addition to our work on *Hyperproof*, we have been working on developing such a touchstone. There are several challenges to be met. One has to do with simply coming up with a framework that does not presuppose that information is presented linguistically, or, for that matter, in any particular medium. Thus, we need a notion of information and information containment that is independent of the form of representation. Given such a notion, we need to be able to say what it means for a given attempt at a proof to be a genuine proof. And we need a way to say all this without knowing in advance the form in which the information is going to be represented. Our first attempts along these lines are spelled out in the companion article mentioned earlier.

Conclusions

In the picture of inference that emerges from traditional logic, the vast majority of valid pieces of reasoning — perhaps all — take place in language, and this sort of reasoning is thought to be well understood. We think this picture is wrong on both counts. First, it is wrong in regard to the frequency with which nonlinguistic forms of information are used in reasoning. There are good reasons to suppose that much, if not most, reasoning makes use of some form of visual representation. Second, it is wrong in regard to the extent to which even linguistic reasoning is accounted for by our current theories of inference. As research in semantics over the past twenty years has shown, human languages are infinitely richer and more subtle than the formal languages for which we have anything like a complete account of inference.

When one takes seriously the variety of ways in which information can be presented and manipulated, the task of giving a general account of valid reasoning is, to say the least, daunting. Nevertheless, we think it is important for logicians to broaden their outlook beyond linguistically presented information. As the computer gives us ever richer tools for representing information, we must begin to study the logical aspects of reasoning that uses nonlinguistic forms of representation. In this way we can hope to do something analogous to what Frege and his followers did for reasoning based on linguistic information. Frege made great strides in studying linguistically based inference by carving out a simple, formal language and investigating the deductive relationships among its sentences. Our hope is that the tools we have begun to develop will allow something similar to be done with information presented in more than one mode.

Comparing Visual and Linguistic Representations

Looking at our theory and examples, we can summarize some of the ways in which diagrammatic representations differ from linguistic representations.

Closure under Constraints

Diagrams are physical situations. They must be, since we can see them. As such, they obey their own set of constraints. In our example from *Hyperproof*, when we represent tetrahedron *a* as large, a host of other facts are thereby supported by the diagram. By choosing a representational scheme appropriately, so that the constraints on the diagrams have a good match with the constraints on the described situation, the diagram can generate a lot of information that the user never need infer. Rather, the user can simply read off facts from the diagram as needed. This situation is in stark contrast to sentential inference, where even the most trivial consequence needs to be inferred explicitly.

Conjunctive vs. Disjunctive Information

It is often said that a picture is worth a thousand words. The truth behind this saying is that even a relatively simple picture or diagram can support countless facts, facts that can be read off the diagram. Thus a diagram can represent in a compact form what would take countless sentences to express. By contrast, sentences come into their own where one needs to describe a variety of mutually incompatible possibilities. For instance, in our *Hyperproof* example, the facts expressed by the various sentences could not themselves be represented in our first diagram. It is only after we have used other information that we can take advantage of this information in subsequent diagrams.

Homomorphic vs. Non-Homomorphic Representation

Another advantage of diagrams, related to the first two, is that a good diagram is isomorphic, or at least homomorphic, to the situation it represents, at least along certain crucial dimensions. In Example 5, there is a mapping of certain facts about our two-dimensional diagrams to corresponding facts about the three-dimensional arrangement of chairs and people. In our Venn diagram proof, there is a homomorphic relationship between regions and sets. This homomorphic relationship is what makes the one a picture, or diagram, of the other. By contrast, the relationship between the linguistic structure of a sentence and that of its content is far more complex. It is certainly nothing like a homomorphism in any obvious way.

Symmetry Arguments

As we have looked at the use of diagrams in solving various sorts of puzzles, we have been struck by the fact that in most such solutions, there is some sort of symmetry argument, either explicit or implicit. We have seen such an argument in the case of Example 5. Such arguments often result in a drastic lowering of inferential overhead, by cutting down on the number of cases that need to be explicitly considered. Diagrams often make these symmetry arguments quite transparent.

There are various ways to try to analyze this phenomenon. For instance, in Example 5, we can think of each diagram as admitting two distinct readings, and so as representing two distinct cases. Whether or not this is the best analysis of the phenomenon, it seems clear to us that the ability to shortcut the number of cases that need to be considered by appealing to symmetries in the diagrams and the situations they represent is a significant aspect of the power of diagrams in reasoning.

Perceptual Inference

Strictly speaking, the eyes are part of the brain. From the cognitive point of view, however, the perceptual apparatus is fairly autonomous. The sorts of things it does are fairly well insulated from the forms of reasoning we engage in at a conscious level. Nevertheless, the perceptual system is an enormously powerful system and carries out a great deal of what one would want to call inference, and which has indeed been called perceptual inference.

It is not surprising that this is so, given the fact that visual situations satisfy their own family of constraints. Indeed, it is the fact that visual situations do satisfy constraints that has made it possible for our perceptual system to evolve in the way that it has. And so it is not surprising that people use the tools this system provides in reasoning. Indeed, it would be incredible were this not so. Once the ubiquity of visual reasoning is recognized, it would seem odd in the extreme to maintain the myth that it is not used in mathematics.

Two Final Remarks

We end by reminding the reader that we are not advocating that mathematics replace linguistic modes of representation with diagrams and pictures. Both forms of representation have their place. Nor are we advocating that logical proofs should be anything less than rigorous. Rather, we are advocating a re-evaluation of the doctrine that diagrams and other forms of visual representation are unwelcome guests in rigorous proofs.

As we have noted, our understanding of valid inference using linguistic modes of representation is far less advanced than is commonly supposed, since the language mathematicians use is far richer than that of first-order logic. This limitation has not blocked progress in mathematics. A full understanding of the power and pitfalls of visual representations is no doubt a long way off. But lack of understanding should not block their use in cases where it is clearly legitimate. Only time and dedication will provide anything approaching a full understanding of the power of visual representation in logic, mathematics, and other forms of human reasoning. But a major impetus for such work can come from the power the computer gives us to use graphical representations in our proofs. For once such tools are in wide use, we logicians will be forced to admit them into our model of mathematical reasoning.

Notes

[1] More precisely, given any three sets A, B, and C we map the various regions in our diagrams to the obvious sets. This is a homomorphism relative to the operations mentioned, so any positive statement formulated in terms of adding and intersecting regions that is true of the diagram corresponds to a true statement about the sets.

2 We note that there are actually two sorts of diagrams in category theory, diagrams as physical parts of proofs, and abstract objects called diagrams used to model them.

3 The one form of uncertainty that our program does not incorporate is uncertainty about the number of objects in the world in question. We allow the user to assume that he can see every object in the world.

References

[1] Barwise, Jon, and John Etchemendy. "Information, Infons, and Inference." In *Situation Theory and Its Applications, I*, Robin Cooper, Kuniaki Mukai, and John Perry, eds., University of Chicago Press, 1990.

[2] Kosslyn, S. M. *Image and Mind*. Harvard University Press, 1980.

[3] Mac Lane, Saunders, *Categories for the Working Mathematician*, Springer-Verlag, 1971.

[4] Stenning, Keith. "On remembering how to get there: how we might want something like a map." In *Cognitive Psychology and Instruction*, ed. by A. M. Lesgold, J. W. Pellegrino, S. W. Fokkema, and R. Glaser, 1977.

[5] Stigler, James. " 'Mental Abacus': The effect of abacus training on Chinese children's mental arithmetic." *Cognitive Psychology* 16(1984), 145-176.

[6] Tennant, Neil, "The Withering Away of Formal Semantics," *Mind and Language* 1(1986), 302-318.

Acknowledgements

The research reported here was supported by an award from the System Development Foundation to the Center for the Study of Language and Information, by the Center for Advanced Study in Behavioral Sciences, and by National Science Foundation grant BNS87-00864. We would also like to thank readers of earlier drafts, especially David Israel, for helpful comments on ideas expressed here.

On the Reluctance to Visualize in Mathematics

Theodore Eisenberg and Tommy Dreyfus

Although the benefits of visualizing mathematical concepts are often advocated, many students are reluctant to accept them; they prefer algorithmic over visual thinking. After establishing the existence of this reluctance, several reasons for it will be discussed. One of these is that thinking visually makes higher cognitive demands than thinking algorithmically, and thus it is quite natural for students to gravitate away from visual thinking. The paper concludes with the discussion of several recent curricular efforts to reverse this trend.

Figure 1 shows a test containing 10 problems. If graduates of your beginning calculus courses are like those at the vast majority of other universities, they will fail this test.[1] (To see just how badly they will do on certain questions, see Mundy [18], Selden, Mason, and Selden [26] and Tufte [33]). Each of the above problems rests firmly upon a concept which has a visual interpretation, and it is our contention that students (and oftentimes their teachers) cannot do many of the problems in this list because they have not learned to exploit the visual representations associated with the concepts.

The list could be extended tenfold, but listing problems doesn't really get at the heart of the matter, for such lists generally focus on specific characteristics of the learning difficulties. Indeed, with respect to problem nine, one typical calculus teacher (who also happens to have authored a calculus textbook) wrote: $f'(-a) = (f(-a))' = (-f(a))' = -f'(a)$. He justified this chain of equalities twice, and it is not even clear that the teacher understood the mistake once it was pointed out! Obviously, this teacher had not focused on a visual graphical image of an odd function when answering the problem. But that is too simplistic an explanation, for he could certainly discuss with authority graphical aspects of odd functions. The point is that the visual characteristics of the problem were not even considered. And that seems to be exactly the problem; visual aspects of a concept are rated secondary to the concept itself. Odd functions are those for which $f(-x) = -f(x)$, and it just so happens that their graphs are symmetric with respect to the origin; monotonically increasing functions are those for which $f'(x) > 0$, and it just so happens that the graph climbs, etc. Although the visual interpretations are often known to students and their teachers, oftentimes they are not in the core of this knowledge and are not

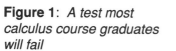

Figure 1: *A test most calculus course graduates will fail*

1. Suppose line L is tangent to the curve $y = f(x)$ at the point (5,3) as indicated below. Find $f(5)$ and $f'(5)$.

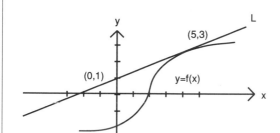

2. Why is:
$$\int_{-1}^{1} \frac{dx}{x^2} = \int_{-1}^{1} x^{-2} dx = \frac{x^{-1}}{-1}\Big|_{-1}^{1}$$
$$= \frac{-1}{x}\Big|_{-1}^{1} = \frac{-1}{1} - \frac{-1}{-1} = -2$$

   obviously wrong?

3. Find the maximum slope of the graph of $y = -x^3 + 3x^2 + 9x - 27$.

4. Evaluate: $\int_{-3}^{3} |x + 2| dx$

5. Using the graph of
$dy/dx = f'(x) = (x-1)(x-2)^2(x-3)^3$,
sketch a graph of $y = f(x)$.

   continued

Theodore Eisenberg is Professor of Mathematics and Education at Ben-Gurion University in Beer Sheva, Israel. He is interested in mathematical concept formation and in ways to think about mathematical concepts in visual formats.

Tommy Dreyfus is Associate Professor of Science and Mathematics Teaching at the Center for Technological Education (affiliated with Tel Aviv University) in Holon, Israel, and is also a consultant at the Department of Science Teaching of the Weizmann Institute of Science in Rehovot, Israel. His main research interest concerns the effects of computer use on students' mathematical thinking patterns.

Figure 1: *continued*

6. If *f* is an odd function on [-a,a], evaluate $\int_{-a}^{a}(b+(f(x))dx$

7. Let $f(x) = \begin{cases} ax, & x \le 1 \\ bx^2 + x + 1, & x > 1 \end{cases}$

   Find *a* and *b* so that *f* is differentiable at 1.

8. In the diagram P, Q and R are points on the graph of *f*. For all *x*, -1< *x* <0, *f"* (*x*) < 0 and for all *x*, 0< *x* <1, *f"* (*x*) > 0. If

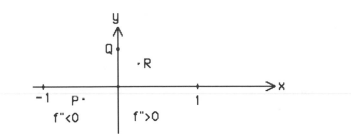

   the derivative of *f* in [-1,1] exists, which of the following must be correct?
   A) *f'* (0) = 0.
   B) *f* has a maximum at *x* = 0.
   C) *f* has a minimum at *x* = 0.
   D) There is a number *c*, -1< *c* <0 at which *f* has a maximum.
   E) There is a number *d*, 0< *d* <1 at which *f* has a maximum.

9. Given: *f* is a differentiable function such that *f* (-x) = -*f* (x). Then, for any given a

   A) *f'* (-a) = -*f'* (-a).
   B) *f'* (-a) = *f'* (a).
   C) *f'* (-a) = -*f'* (a).
   D) None of the above.

10. Suppose *f* is a continuous function. Which of the following is correct?

   A) $\int_{a}^{b}f(x-k)dx = \int_{a-k}^{b-k}f(x)dx$

   B) $\int_{a}^{b}f(x-k)dx = \int_{a}^{b}f(x+k)dx$

   C) $\int_{a}^{b}f(x-k)dx = \int_{a+k}^{b+k}f(x)dx$

   D) $\int_{a}^{b}f(x-k)dx = \int_{a+k}^{b+k}f(x+k)dx$

   E) None of the above.

exploited in thinking processes. In order to understand why this is the case, carefully constructed and coordinated studies must be done, but this is not as simple as it sounds.

Background. Whenever possible, students seem to choose a symbolic framework to process mathematical information rather than a visual one (Eisenberg and Dreyfus [7]). While this is also true for many teachers, it does not seem to hold for professional mathematicians. For them, the choice of representation in which to solve a problem seems to depend as much on the problem itself as on personal preferences. An informal survey of our colleagues underscored this point. We tried to assess their views on visualization, specifically to determine whether or not they themselves visualize. Although some stated that they see everything in terms of pictures, others said that they see nothing in pictures. Indeed, one stated that she hates teaching beginning calculus because she cannot visualize many of the problems that have to be solved. Algebraists expressed more personal reluctance to visualize than analysts, and females more than males. Mundy [19] has also observed differential performances between the sexes on spatial visualization ability in calculus.

On the surface, it thus appears as if visual processing is more difficult for students than analytic processing. Although the reasons for this are far from clear, we will try to elucidate them in this paper. In order to be specific, we will concentrate on examples from calculus, although much of the general reasoning and underlying phenomena also occur in other domains of mathematics.

As an example, consider the work by Selden, Mason, and Selden [26]. They gave their calculus graduates five questions they thought everyone passing calculus should be able to answer. These questions had not been taught in the course, yet each could be answered easily if the student had a visual understanding of the derivative. Question 7 of Figure 1 is taken from their questionnaire; their other questions were of a similar level. Their findings were most discouraging: "Not one student got an entire problem correct.

Most couldn't do anything" (p.48). The authors then went on to discuss the Washington D.C. colloquium on *Calculus for a New Century* [28]. Many technical reasons were posited there as the cause of student failures. These included large lecture sections, use of inexperienced teachers and teaching assistants, poorly prepared students, the lack of proper placement tests, etc. But the researchers claimed, "Our students suffered none of these handicaps, yet came away from the course with little knowledge" (p.49).

Clements [3] has observed that the tendency to avoid visualization exists even in the mathematically precocious. Terence Tao is a mathematically precocious Australian. At the age of 8 he scored 720 on the SAT-M exam; he felt at ease in discussing calculus and abstract algebra and he published his first paper on perfect numbers. At age 10 he gained university acceptance. Clements has studied Terence Tao in an effort to determine his thinking processes. In this study, Tao was given batteries of exams and subjected to many interviews. When focusing on this work with respect to visualization abilities, Clements concludes: "Analysis of (his) methods ... strongly suggested that he preferred to use non-visual, analytic methods whenever these occurred to him, even if they required more complicated thinking than more visual methods which could be used ... While he has well developed spatial ability, when attempting to solve mathematical problems, he has a distinct, though not conscious, preference for using verbal-logical, as opposed to visual thinking" (p.225).

Mundy [18] found that students often have only a mechanical understanding of basic calculus concepts. She concluded that this happens because students haven't achieved a visual understanding of basic underlying notions. This thesis, that students master only mechanics, seems to have become a *cause célèbre* for reforming the calculus, as evidenced by the plethora of recent articles on this subject in MAA publications (see, e.g., Douglas [6], Gillman [9], Ralston [21], and Steen [27]). But the authors of such articles seem to offer only anecdotal evidence on this crucial issue.

Although most instructors can list specific problems and clusters of problems which cause students difficulty, few have taken the time to analyze why these difficulties occur, nor have they thought in a serious way about how they can be rectified. It is a fact that at least 60% of all beginning calculus students either fail or drop the course, but this statistic is meaningless without attempting to analyze why it occurs. Although Mundy has observed the obvious, that students have only a mechanical understanding of basic notions, she and others have tried to delve deeper into the issue. In research articles they try to understand the reasons why students do not succeed with basic collegiate mathematics. They try to analyze learning problems from both sociological and cognitive points of view. Some of these studies will be discussed in the next section.

Many believe that this lack in understanding is to a large extent due to the failure to establish explicit and detailed connections between the visual and analytic aspects of mathematical concepts and procedures. Their studies have tried to document this belief and how learning problems might be eradicated by stressing a visual approach to mathematical concepts.

Some Research Results

Mundy [18] asked 973 calculus course graduates to evaluate

$$\int_{-3}^{3} |x+2|\,dx \ .$$

It was given as a multiple-choice question with choices 0, 9,12,13, and 14. Only 5.4% of the students correctly answered this question. Twenty-four percent gave 0 as the answer, 22% said it was 9, and 48% answered 12. Mundy concluded that the students did not have a visual understanding that integrals of positively valued functions can be thought of in terms of an area under a curve. Yet these same students had no trouble computing the area between two "nice" intersecting curves.

Figure 2: *A Pointwise and Across-Time Question*

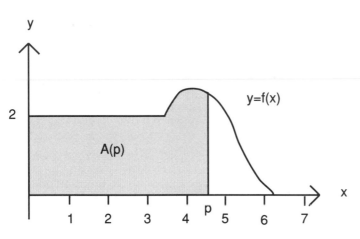

Dick [5] focused on Mundy's findings and built a study assessing whether or not students who could correctly draw the graph of such functions would use the graph to monitor symbolic calculations. His subjects were students near the end of a semester course on probability and statistics. He gave them the function $f(x) = |x - 2|$ when $1 < x < 3$ and 0 elsewhere. He asked them to a) graph $y = f(x)$ and b) find the Prob($X > 3/2$), where X is a continuous random variable with probability density function given by $f(x)$. He examined three parts of their work: 1) the correctness of the graph of $f(x)$, 2) the formulation of Prob($X > 3/2$) as

$$\int_{3/2}^{3} |x-2|\,dx$$

and 3) the computation of Prob($X > 3/2$) as coming from either direct analysis of the graph or from computing

$$\int_{3/2}^{2} -(x-2)\,dx + \int_{2}^{3} (x-2)\,dx \ .$$

Ninety-two percent correctly graphed the probability density function and 89% also gave a correct formulation of Prob($X > 3/2$), but of these only 44% correctly computed the integral (72% by integration and 28% by direct graphical analysis). With respect to the incorrect answers, Dick [5] concluded: "There was no evidence of graphical interpretation of any kind ... even college students of relatively advanced mathematical training can be expected to ignore the use of their own graphs, even when these are produced immediately preceding a computational problem for which they could be used" (p.2).

Monk [17] has made similar episodic observations that students lack a visual image of basic notions in elementary calculus, but he believes that he has pinpointed a specific learning obstacle. He claims that although students do not have trouble with understanding graphically represented data in a

Figure 3: *A Typical Graph of the Speed of a Golf Ball*

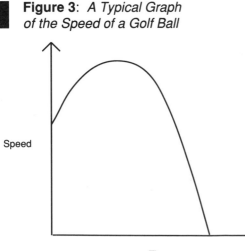

"pointwise" manner, serious difficulties do arise in developing an "across-time" understanding which is exactly the kind necessary in calculus. Figure 2 illustrates this problem:

> Pointwise: Determine the values of $A(1)$ and $A(3)$.
> Across-time: The point p moves from 4.5 to 6.0. Does the area $A(p)$ increase or decrease?

On this and similar problems, 85% of the calculus course graduates could correctly answer pointwise problems, but only 53% correctly answered across-time questions. Swan [29] and his colleagues have also documented that students cannot answer across-time questions. They have shown that if one asks a student to graph the speed of a golf ball being hit off a tee on a mountain top, one will most likely get something completely wrong like that in Figure 3. Monk and Swan conclude that beginning calculus courses are very lax in helping students develop the skills to produce, analyze, and use graphs in an across-time manner.

Vinner [34] developed a calculus course around the hypothesis that "among those who succeed in mathematics, the algebraic mode is more common when solving routine or almost routine problems" (p.149). In order to find out to what extent visual considerations can become a natural part of college students' mathematical thinking, he developed a course which emphasized the visual aspects of calculus. "Every algebraic concept which can be approached from a visual starting point was approached so (continuity, derivative, increasing and decreasing of functions, convexity, local minima and maxima, definite integral, etc.) In addition, the visual meaning of every theorem that has a visual meaning was presented to the students and very often visual considerations were given as proofs for the theorems" (p.150). On a questionnaire he asked students to reproduce the proof that if $y = f(x)$ is a positive, strictly increasing function on $[a,b]$, then

$$(b-a)f(a) < \int_a^b f(x)dx < (b-a)f(b).$$

He also asked them to formulate and to prove the mean value theorem. In the lecture, proofs for these statements had been given in both visual and algebraic contexts. In both questions, students were explicitly asked to make a suitable drawing. In spite of this the results showed a definite bias toward an algebraic approach.

Summary

The work of Mundy, Dick, Monk, Swan, and Vinner supports the preliminary finding by the authors that students have a strong tendency to think algebraically rather than visually. Moreover, this is so even if they are explicitly and forcefully pushed towards visual processing. It has been claimed that many learning difficulties, at least in the calculus, could be alleviated or even avoided if students were brought to internalize the visual connotations of calculus concepts. But why does this not happen?   And why, if the visual connotations are there, are the links between the visual and the analytic aspects of the concepts and procedures lacking? What is needed to generate a visual understanding which would include the ability to solve problems utilizing both visual and analytic thinking in concordance with one another, and help students feel at ease in both domains?

Reasons for Avoidance

Students' preference for making nonvisual arguments is not accidental. In this section, some possible reasons for this will be discussed. In order to be more concrete, let us start with an example of what is meant by a visual versus nonvisual explanation.

An example. Assume that you want to show your calculus class that if F and G are differentiable functions inverse to each other and $F(a) = b$ then $F'(a) = 1/G'(b)$. A nonvisual way to achieve this is to start from $G(F(x)) = x$ and use the chain rule to obtain $G'(F(x))F'(x) = 1$, from which the result follows by substituting a for x and b for $F(x)$. A visual way to show the result would be to use the graph in Figure 4 (which could be on a slide or drawn by a computer), and reach the conclusion from the three facts that $F'(a)$ is the slope of K, $G'(b)$ is the slope of L, and K and L are symmetric to each other with respect to $y = x$.

Figure 4: *A visual explanation of* F'(a) = 1/G'(b)

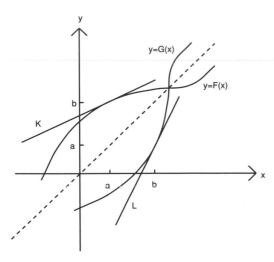

There are considerable differences between the two arguments, in both the prerequisite knowledge they demand as well as in the understanding that is likely to follow from internalizing the argument. The analytic argument is short, neat, assumes little and gives the result without any of its wider implications. It is easy for the student to learn, easy to apply to exercises requiring exactly the computation of the derivative of an inverse function at a point, but unlikely to enable the student to achieve more than mechanically doing exactly this. It is also easy to teach, does not require the preparation of a graph, slide, or computer program, and separates the argument completely from other arguments which may be similar. It corresponds to what students expect from a mathematical proof and will not usually lead to a discussion.

The visual argument, on the other hand, builds on prerequisite visual knowledge such as that the derivative is the slope of the tangent line, and the product of the slopes of lines symmetric with respect to $y = x$ equals 1. It shows additional, related, but not essential, information such as the relative location of the points (a,b) and (b,a), as well as their significance. It is thus more difficult to understand and it is likely to give rise to a discussion in class. We leave it up to the reader to estimate which of the arguments would produce better results on test questions such as: If $f(x) = x^5 + 8x^3 + 4x - 2$, $0 \le x < \infty$, compute $(f^{-1})'(-2)$ or, if $h(x) = x^3 + 2x$, $0 \le x < \infty$, and if g is the inverse function of h, compute $g'(3)$.

Below we relate in more detail three of the differences pointed out above: a cognitive one (visual is more difficult), a sociological one (visual is harder to teach) and one related to beliefs about the nature of mathematics (visual is not mathematical).

Beliefs About the Nature of Mathematics

It has evolved that nonvisual frameworks are used to communicate mathematical ideas. This custom is based on the belief of many mathematicians, teachers, and students that mathematics is nonvisual, regardless of whether or not a visual representation is at the base of an idea. We in the academic community have only ourselves to blame for perpetuating this view of mathematics.

This belief is deeply grounded in us, even among many who advocate visualization. Consider Vinner's mindset on this: "We do believe that in more advanced courses (like analysis) the algebraic approach should replace the visual approach" (p.155). This implies that he considers higher mathematics as being non-visual, but then three sentences later he recommends: "The belief that a visual proof is not a mathematical proof should be eliminated" (p.156). The idea that higher mathematics must be communicated in a nonvisual framework is shared by many in the mathematics community, and it is deeply rooted. For example, consider Hilbert's comments as stated in Hadamard's text [11]: "I have given a simplified proof of part (a) of Jordan's theorem. Of course, my proof is completely arithmetizable (otherwise it would be considered non-existent); but, investigating it, I never ceased thinking of the diagram (only thinking of a very twisted curve), and so do I still when remembering it" (p.103).

Figure 5: *Proofs without words*

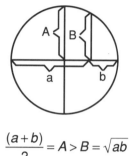

$$\frac{(a+b)}{2} = A > B = \sqrt{ab}$$

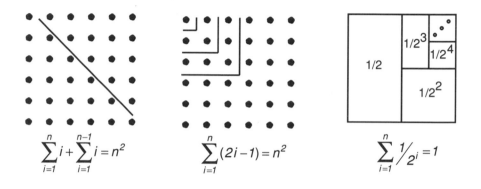

$$\sum_{i=1}^{n} i + \sum_{i=1}^{n-1} i = n^2 \qquad \sum_{i=1}^{n}(2i-1) = n^2 \qquad \sum_{i=1}^{n} \frac{1}{2^i} = 1$$

From this it seems clear that Hilbert did not consider the visual argument sufficient for being a proof. The proof had to be arithmetizable, otherwise it did not exist. This raises all sorts of questions as to the nature of mathematics. Davis and Hersh [4] have also addressed this problem. Specifically, they have written about the roles played by symbolization and formalization in defining the nature of mathematics. But it seems as though the influence of the Bourbaki school, which emphasizes the analytic expression of mathematical ideas, has been so ingrained into our psyche that we consider it as being the only acceptable way of professional communication. To many in our profession, a proof without words is not a proof!

Over the years proofs without words have appeared in *The College Mathematics Journal, Mathematics Magazine*, the MAA *Monthly*, and various other publications. They are often sandwiched between articles and seem to be used as space fillers, which might say something in and of itself. But some of us have collected these gems and have presented them to students at appropriate times. Several years ago one of us (T.E.) gave a seminar on this topic to university colleagues.[2] The purpose of the seminar was simply to expose colleagues to this visual way of presenting mathematics. It soon became apparent that these diagrams were not considered proofs at all, but something much less in status: mathematical mnemonics. Figure 5 presents four of the diagrams discussed. The consensus was that although the diagrams could be used to generate a proof, they certainly were not proofs in and of themselves and most of the participants were adamant on this point. It is true that one must make a jump to the *n*th case in problems b, c, and d of Figure 5, but the criticisms did not center on this. Specifically, they centered around the idea that there is one and only one way to communicate mathematics, and "proofs without words" are not acceptable.[3]

If visual methods are considered of minor value by mathematicians and teachers of mathematics, if they are at best mathematical mnemonics but by no means allowed as a full-fledged argument, then this attitude will be conveyed by these mathematicians and teachers to students and there is no wonder that students react in the way they do. However, as noted above, most mathematicians do use visual exploration and argumentation in their daily work. Why then do they not pass this use of visual thinking on to their students? Or, if they do, why do the students not adopt it?

A Sociological Rationale

There appears to be a difference between methods of processing information used by mathematicians in their own work, which are very often visual, and those they teach, in which the visual element is relegated to an illustrative role if it is used at all. To understand the origin of this difference, we need to consider Chevallard's [2] theory of mathematical didactics in some detail (see also Seeger, Steinbring and Straesser [25]). The central idea of this theory is that of a "didactical transposition," which is defined as the change knowledge undergoes as it is turned from scientific, academic knowledge to instructional knowledge as taught in school. In other words, Chevallard describes how knowledge enters schools and the changes it undergoes as it does so. This didactical preparation of knowledge has several aspects, the most important of which, in our context, are linearization and compartmentalization.

Academic knowledge is very intricate and contains many links and connections; these cannot be presented as a package since their presentation must be sequential. The elements of this knowledge must thus be taken apart and ordered sequentially. This process necessarily destroys many of the links and therefore a considerable part of the unity of the knowledge. The didactical preparation of knowledge implies the formulation of a linear text which structures the knowledge, giving it for instance, a beginning and an end. This is not the way knowledge "sits in the mind of the mathematician." Compartmentalization of knowledge occurs not only because links between concepts and procedures are destroyed or omitted, but also because in school, knowledge is necessarily taught separated from its context. Knowledge is thus split from a "body of knowledge" into a large number of isolated "bits of knowledge." Moreover, prerequisite knowledge must be made explicit, rather than being presented implicitly. Knowledge is presented in such a way as to be as similar as possible to knowledge already possessed by the learner. The degree of novelty must be kept artificially low, and often this is achieved by means of building algorithms and procedures of that which has to be taught. New knowledge is formulated, for the purpose of teaching, in a way that stresses computational procedures, and this necessarily allows for sequential presentations.

Chevallard's theory has an immediate application to the question of visual versus analytic processing of information. Analytic processing uses sentential representations, in which the information forms a sequence of expressions; visual processing uses diagrammatic representations. Larkin and Simon [15] claim that the processing of these two types of representations is necessarily different. "The fundamental difference between our diagrammatic and sentential representation is that the diagrammatic representation preserves explicitly the information about the topological and geometric relations among the components of the problem, while the sentential representation does not" (p. 66). Chevallard claims that school knowledge is necessarily sequential; it is thus best represented sententially, not diagrammatically. As a consequence, teachers prefer to use sentential rather than diagrammatical representations. Sentential representations are those which are appropriate for knowledge having undergone the didactical transposition, diagrammatical ones are not. Because school mathematics is usually linearized and algorithmetized, it is the most efficient way for teachers to present material to students. In other words, it makes perfect sense that students choose analytic rather than visual procedures.

A Cognitive Rationale

Another way to look at this situation is to build upon the idea of cognitive efficiency. It is relatively easy to present an argument that has already been ordered in a linear sequential manner. It is much more difficult to present a two-dimensional array of information with multiple links between various pieces of information and with implications in many different directions. A diagram is a very complex and concentrated collection of information. If the entire information contained in a diagram is reformulated in sentential form, it will take a lot of space. Even if two representations of a mathematical concept contain the same information, much of this information will be explicit in the diagram but implicit in the analytic representation. Consider, for example, the function whose analytic representation is $f(x) = |x - 2| - 2$. From a graph of this function we can see explicitly that it is not differentiable at $x = 2$ (rather than $x = -2$), that it has zeros at $x = 0$ and $x = 4$, that it decreases at $x = 1$, and so on. All this information is contained in the formula for the function as well, but it is implicit. When we need it for solving problems we must first dig it out. It follows that diagrammatic information is often more useful. This added usefulness is not because there is more information in the diagram but because this information is presented in a way that explicitly exhibits important pieces of information and important conceptual links between them. But now to the other side of this coin: diagrammatic representations are not immediately intelligible to the uninitiated. It takes cognitive processing to make sense of diagrammatic representations. Or, to cite again Larkin and Simon [15]: "Diagrams are useful only to those who know the appropriate computational processes for taking advantage of them. Furthermore, a problem solver often also needs the knowledge of how to construct a 'good' diagram that lets him take advantage of the virtues..." (p. 99). It is no wonder that students who have not been taught appropriately refuse to draw diagrams or graphs and do not know how to take advantage of those provided to them or even those they have drawn themselves. It simply means that they have not had the opportunity to acquire the difficult skill of interpreting such diagrams, of using them in solving problems and, even less, of constructing useful diagrams themselves.

Although Janvier [14] and Goldenberg [10] have researched the conceptual difficulties involved in reading and interpreting Cartesian graphs, there is overall a dearth of studies in this area. The mathematics community does not know much beyond the fact that using diagrams in class seems to be an insufficient way to produce desired results. Our efforts to date are still inadequate for getting students to translate back and forth between visual and analytic representations of the same situation, although as a result of new efforts in this area, this situation may soon change.

Recent Efforts

The above sections have documented that it is not common for students to process mathematics visually. There seem to be many explanations as to why this is so. But the mathematical community is well aware of the benefits of having visual concept images of mathematical ideas.

The Authorities

Halmos [12], speaking of Solomon Lefschetz (editor of the *Annals*), stated: "He saw mathematics not as logic but as pictures" (p. 87). Speaking of what it takes to be a mathematician, he stated: "To be a scholar of mathematics you must be born with ... the ability to visualize" (p. 400) and most teachers try to develop this ability in their students. Pólya's [20] "Draw a figure ..." (p. xvi) is classic pedagogical advice, and Einstein and Poincaré's views that we should use our visual intuitions are also well known. The list of authorities advocating visualization could be extended tenfold. We dare say that no mathematician would denigrate the benefits of visualization, although they may not laud it uniformly. But it is only recently that a concerted effort seems to be underway to bring visualization into the school and collegiate curriculum. This volume is one such testimony. A few others from around the world are described below. Lowenthal and Vandeputte [16] discuss efforts in Belgium to help secondary school students develop the skill of interpreting functions with little or no use being made of a computer. The program uses the idea of manipulating a carefully developed sequence of stock functions. For example, tenth grade students are led to graph $y = \cos(10x) + \cos(8x)$ by considering the inequality $-2\cos(x) \le 2\cos(x)\cos(9x) \le 2\cos(x)$, which suggests that the requested graph must lie between those of $y_1 = -2\cos(x)$ and $y_2 = 2\cos(x)$. They report success in getting average ability students to have an intuitive understanding of most textbook functions.

Curricular Efforts

Flanders, Korfhage, and Price [8] have written an entire calculus text built upon the idea of visual augmentation. More recently several projects have taken a similar approach integrating the use of computers. In these curricula, visualization, or more precisely the bridges between visual and analytic representations of the same mathematical concept, are stressed. See, for example, the programs of Tall [30] [31] in the United Kingdom, Artigue [1] in France, Heid [13] in the United States, Schwarz [24] in Israel and Ruthven [23] in the UK.

Tall's graphic calculus is based on a cognitive approach to the curriculum. It uses software that is specifically designed to enable the user to manipulate generic examples of mathematical concepts. That is, it helps students interpret situations based upon visual intuitions. The learner is directed through a suitable sequence of activities with examples and non-examples towards the fundamental properties of the concept. Sample concepts are the slope of a graph (which exhibits "local straightness" through magnification), the area under a curve (which illustrates the sign of the area using dynamic growth), and the solution of a differential equation (where students see it develop graphically in real-time computer simulation).

Artigue also builds her curriculum on suitable computer software which is used both in graphical and numerical mode. It uses the computer as a tool to help the student develop a qualitative approach to differential equations. The role of the computer is essential here, because it reduces the complexity of the problems without losing the global picture associated with them. It is used in the graphical mode to search for curves compatible with a given slope field, and then the numerical mode is brought into play. The entire process is built on a continual process of association between graphs and equations. Graphics and visual information are given a role beyond that of merely being another representation of a problem. They are the central objects from which information is processed both visually and symbolically.

Heid developed a conceptual calculus course in which a graphing and symbol manipulation computer program was used to perform routine manipulations.

Only the last three weeks of the course were spent on skill development. The analysis of applications through graphs was a fundamental ingredient of her course. Overall, her students showed better understanding of course concepts, and they performed almost as well on a final exam of routine skills as a class of students who had practiced these skills for the entire semester.

Schwarz has developed an introductory functions curriculum based on a computer environment called Triple Representation Model (TRM). This problem-based curriculum has been specifically designed to focus attention on within-representation and between-representation relationships. The student learns by operating in the algebraic, graphical and tabular representations, and by measuring and comparing the effect of an operation in various representations. For example, a good strategy for finding the solution to the equation $f(x) = 7$ with $f(x) = x^3 - 4x + 1$ is first to build a table of values which roughly locates the solution between 2 and 3, then to draw the graph with the corresponding bounds ($2 < x < 3$, $6 < y < 8$), and finally to compute the solution algebraically to within a certain accuracy on the basis of the approximate graphical solution. The TRM curriculum promotes such strategies.

Ruthven studied A-level students in Britain who had continuous access to graphic calculators for at least a year. He found that these students not only improved far more than their colleagues without calculators in achievement, but also that their mathematical approaches and arguments were more likely to be graphical.

Conclusion

In the terms of Rival [22], "Mathematicians are rediscovering the power of pictorial reasoning." But there seem to be valid reasons, as outlined in this article, as to why students have trouble thinking in visual frameworks. Understanding these reasons should help in the development of suitable materials and learning strategies to promote visual thinking. The programs outlined above seem to be a step in the right direction. As a result of the new curricula, it is hoped that students will develop a deeper understanding of the underlying concepts. Let us hope that Selden, Mason, and Selden's comments: "Not one student got an entire problem correct. Most couldn't do anything ... (they) came away from the course with little knowledge" will never be heard again. It is said that Einstein thought in terms of pictures. It seems that it would behoove our students to think that way too.

References

[1] Artigue, M., "Ingenierie didactique a propos d'equations differentielles," *Proceedings*, PME-11 Congress (Vol. III), 1987, Montreal, 236-243.

[2] Chevallard, Y., *La transposition didactique du savoir savant au savoir enseigne.* La Pensee Sauvage, Grenoble, France, 1985.

[3] Clements, M. A., "Terence Tao," *Educational Studies in Mathematics*, 15(1984), 213-238.

[4] Davis, P. J. and R. Hersh, *The Mathematical Experience*, Birkhauser, Boston, 1981.

[5] Dick, T., "Student Use of Graphical Information to Monitor Symbolic Calculations," (Working paper available from author at Oregon State University).

[6] Douglas, D. G. (Ed), *Toward a Lean and Lively Calculus: Report of the Conference/Workshop to Develop Curriculum and Teaching Methods for Calculus at the College Level.* Mathematical Association of America, Washington, D.C., 1987.

[7] Eisenberg, T. and T. Dreyfus, "On Visual Versus Analytical Thinking in Mathematics," *Proceedings*, PME-10 Congress (1986), London, 153-158.

[8] Flanders, H., D. Korfhage, and J. J. Price, *Calculus*. Academic Press, New York, 1970.

[9] Gillman, L., "Two Proposals for Calculus," *Focus*: Newsletter of the MAA 7(1987), 3.

[10] Goldenberg, E. P., "Believing is Seeing: How Preconceptions Influence the Perception of Graphs," *Proceedings*, PME-11 Congress (1987), Montreal, 197-203.

[11] Hadamard, J., *The Psychology of Invention in the Mathematical Field*. Dover, New York, 1954.

[12] Halmos, P. R., *I Want to be a Mathematician*. Mathematical Association of America, Washington, D.C., 1987.

[13] Heid, M. K., "Resequencing Skills and Concepts in Applied Calculus Using the Computer as a Tool," *Journal for Research in Mathematics Education* 19(1988), 3-25.

[14] Janvier, C., *The Interpretation of Complex Cartesian Graphs Representing Situations — Studies and Teaching Experiments*. (Doctoral dissertation). University of Nottingham, UK,1978.

[15] Larkin, J. H. and H. A. Simon, "Why a Diagram is (Sometimes) Worth Ten Thousand Words," *Cognitive Science,* 11(1987), 65-99.

[16] Lowenthal, F. and C. Vandeputte, "Manipulations of Cartesian Graphs: A First Introduction to Analysis," *Focus: On Learning Problems in Mathematics*, 11(1989), 89-97.

[17] Monk, G. S., "Students' Understanding of Functions in Calculus Courses," In *Humanistic Mathematics Network Newsletter* (No. 2, 1988). (Available from A. White (Ed.), Dept. of Math. Harvey Mudd College).

[18] Mundy, J., "Analysis of Errors of First Year Calculus Students," In *Theory, Research and Practice in Mathematics Education*. A. Bell, B. Low and J. Kilpatrick, (Eds.). *Proceedings*, ICME 5. Adelaide, 1984. Working group reports and collected papers. Shell Center, Nottingham, U.K., 170-172.

[19] Mundy, J., "Spatial Training for Calculus Students: Sex Differences in Achievement and in Visualization Ability," *Journal for Research in Mathematics Education*, 18(1987),126-140.

[20] Pólya, G., *How to Solve It*. Princeton University Press, Princeton, NJ, 1973.

[21] Ralston, A., "Will Discrete Mathematics Surpass Calculus in Importance?" *College Mathematics Journal,* 15(1984), 371-373.

[22] Rival, I., "Picture Puzzling: Mathematicians are Rediscovering the Power of Pictorial Reasoning," *The Sciences,* 27(1987), 41-46.

[23] Ruthven, K. *The Influence of Graphic Calculator Use on Translation from Graphic to Symbolic Forms.* (Technical report, 1989. Available from author, Dept. of Education, University of Cambridge, UK.)

[24] Schwarz, B., *The Use of a Microworld to Improve the Concept Image of a Function: the Triple Representation Model Curriculum.* (Doctoral Dissertation, 1989). Weizmann Institute of Science, Rehovot, Israel.

[25] Seeger, F., H. Steinbring, and R. Straesser, "Die Didaktische Transposition," *Mathematica Didactica*, 12(1989), 157-177.

[26] Selden, J., A. Mason, and A. Selden, "Can Average Calculus Students Solve Nonroutine Problems?" *Journal of Mathematical Behavior,* 8(1989), 45-50.

[27] Steen, Lynn A., "Taking Calculus Seriously," *Focus*: Newsletter of the MAA, 6(1986), 4.

[28] Steen, Lynn A., *Calculus for a New Century: A Pump, Not a Filter*. (MAA Notes No. 8) Mathematical Association of America, Washington, D.C., 1987.

[29] Swan, M., "On Reading Graphs," Talk at ICME-6, Budapest, Hungary, 1988. (Available from author, Shell Center,University of Nottingham, UK.)

[30] Tall, D. O., *Graphic Calculus* (software), Glentop Press, London, 1985.

[31] Tall, D. O., *Building and Testing a Cognitive Approach to the Calculus Using Interactive Computer Graphics*. (Doctoral Dissertation, 1986). University of Warwick, U.K.

[32] Thomas, G. B. and R. L. Finney, *Calculus and Analytic Geometry* (7th ed.), Addison-Wesley, Reading, MA, 1988.

[33] Tufte, F. W., "Conceptual vs. Procedural Knowledge in Introductory Calculus: Programming Effects," *Proceedings*, Conference on Technology in Collegiate Mathematics (1988), Ohio State University, Columbus, Ohio.

[34] Vinner, S., "The Avoidance of Visual Considerations in Calculus Students," *Focus: On Learning Problems in Mathematics*, 11(1989), 149-156.

Notes

[1] Although there are many levels of university calculus, throughout this paper we are speaking about the level of course found in Thomas and Finney's text [32].

[2,3] Tel Aviv University, Winter term, 1987. Several of the problems presented in that seminar were also presented at the International Congress of Mathematical Education (ICME-6) in Budapest, Spring, 1988. The same sort of criticisms were heard.

Seeing Beauty in Mathematics: Using Fractal Geometry to Build a Spirit of Mathematical Inquiry

E. Paul Goldenberg

This paper describes in some detail a vision for mathematics education in grades 7 through 12 and some of the practical considerations that my colleagues and I at EDC are taking into account in our endeavor to reify that vision. We bring two concerns to our focus on mathematics education: doing mathematics, and mathematics worth doing.

Adopting a visual and experimental style of mathematical inquiry and learning, we seek to demonstrate that it is possible to make dramatic and fundamental changes in students' engagement in mathematics — to foster a spirit of self-propelled inquiry, mathematical generativity, and strong personal interest in developing competence in intellectually challenging traditional and contemporary mathematical domains. As a test case, we have chosen fractal geometry, a field that represents important frontier mathematical research and a valuable tool in the sciences, and that can, for these and other reasons, play a pivotal role in the curriculum in tomorrow's schools.

These goals raise broad questions. How should the transfer of research mathematics to the high-school curriculum proceed? What are the most promising topics and approaches for introducing this particular new mathematics into the curriculum? What parts are appropriate for grades 7 to 12? How can fractal geometry best be used as a powerful exemplar of a visual and experimental approach to mathematical thinking? How might the fractal content and the visual, experimental approach broaden and deepen students' mathematical thinking and their interests in and perceptions of mathematics? Can the study of fractal geometry in a suitably designed computer-based learning environment help students experience mathematics more the way mathematicians do and enhance student understanding of other mathematical and scientific pursuits? To address these questions, our project is crafting, investigating, and demonstrating a visual and experimental approach to mathematical thinking that offers a form of mathematical meaning-making that purely symbolic approaches often do not. We use the visual beauty of fractals, and the remarkable ways in which they develop, support, and defy our intuitions, to attract students to the *intellectual beauty of the mathematics* behind the fractals. We use the newness of the field, aided by high speed computer graphics tools, to help students explore and create mathematics by asking genuinely new questions and performing actual mathematical research.

E. Paul Goldenberg received his doctorate in curriculum and supervision from Harvard Graduate School of Education. He is currently a Senior Scientist at Education Development Center, Inc. (EDC) in Newton, Massachusetts, holds an appointment as Associate in Education at Harvard Graduate School of Education, and is an adjunct faculty member at Lesley College in Cambridge, Massachusetts. He is the author of Special Technology for Special Children, *principal author of* Computers, Education, and Special Needs *(Addison-Wesley), and* Exploring Language with Logo *(The MIT Press), and editor of the MIT Press series* Exploring With Logo. *He directs research in visualization in mathematics and collaborates in the development of curricula in mathematics and linguistics.*

The Experience of Doing What Mathematicians Do

Contributions of Advanced Technology to Mathematics and Mathematics Education

Computer-supported experimentation and high-resolution, high-speed, animated graphics have changed the face of mathematical research. Several contemporary mathematical domains owe their very existence to interactive graphics on the computer. Fractal geometry, for example, has roots extending back nearly a century, but did not develop as a field until recently with the advent of graphical computation. Now it is not only a flourishing and productive field of pure mathematical research, but also a highly prized tool whose insights are applicable in a wide variety of sciences. Research in other domains, such as topology, whose development had already progressed considerably without advanced technologies gained a new dimension when studies previously undertaken almost exclusively through symbolic manipulations were augmented with graphical representations.[1]

The Public Image and the Curriculum

In the words of Douady [8], mathematicians "generally face problems nobody knows how to solve" while students "face problems which they believe their teacher can solve."

The vitality of mathematics, as mathematicians experience it, is in sharp contrast to its public image. As it is commonly perceived, mathematics is the least creative of subjects: a dead, unchanging body of facts and techniques handed down from the ancients and tolerating no room for inquiry, every question bearing one and only one answer, an answer that is already known by someone. Such an image can hardly draw interest.

The perception that mathematics does not change is no doubt partly the result of the resistance of the mathematics curriculum to change. Decade after decade, students are taught largely the same *old* mathematics regardless of new mathematical developments, new scientific needs, and new technological capabilities. There is no justification for such conservatism, and, indeed, no other field would tolerate it. A curriculum suggesting that all mathematical discovery ceased nearly two centuries ago hobbles our ability to attract bright, interested students into mathematics; and it handicaps the thinking of those who do elect to study mathematics and science by depriving them of modern mathematical ideas and tools.

Attracting students to mathematics, or, more precisely, maintaining and justifying their interest in mathematics, seems especially important at junior high school and early high school levels. By tenth grade, many students have stopped taking the subject altogether, leaving no future chance to discover parts of it that might appeal to them and cutting themselves off from opportunities to pursue studies that depend on higher mathematics. Grades 7 to 9 in particular are crucial targets for the amelioration of the conditions that give rise to students' rapid flight from mathematics courses, especially the sharp decline in girls' interest and participation.

But unchanging content is not the only cause for the popular perception that mathematics is dead, perhaps not even the greatest factor. As a colleague is fond of saying, "We spend all of our time teaching students the rules of the game, and we never give them a chance to play it."[2] Far from being rigid, mathematics *can* be the most freeing of subjects, a game in which the player is free to invent any set of playing pieces, rules, and constraints, and then reason out or observe the consequences of these choices. It is a game whose players frequently use the words *elegance* and *beauty*, and whose beauties are both visual and intellectual. Yet we show little to none of this beauty to our students.

Further, the teaching of both mathematics *and* science is beset by the fragmentation of the traditional curricula and the apparent irrelevance of any one course to any other. As biology, chemistry, physics, and the earth sciences are commonly taught, they appear to share few if any important ideas. Mathematics suffers in a similar way: geometry is almost totally isolated from the thin thread that ties algebra to calculus. The result cannot help but support a notion that learning in mathematics and science consists of acquiring large numbers of unrelated facts and formulas. Without a theme, there can be no variations; without some kind of orderliness, no creativity.

New Opportunities

The opportunity to perform mathematical experiments using interactive visual media can be as valuable to students as to mathematicians. Properly designed computer-supported environments can provide, through their concreteness, a scaffold for reasoning and a matrix for problem posing and can foster the development and use of qualitative, visually-based reasoning styles to augment the traditionally taught symbolic-deductive methods.[3] Beyond affording exciting opportunities for students to engage in new mathematical ways of thinking, such environments can also open up previously inaccessible mathematical domains, allowing precollege students to investigate topics that are often dismissed as too advanced for them.

Of course, new opportunities oblige us at the very least to reconsider our choices. This paper describes in relatively concrete terms a strategy through which we believe it is possible to help students in grades 7 through 12 engage themselves in a visual, experimental mathematics, finding, posing, and attempting to solve problems much as a creative research mathematician would. For reasons that will be given later, fractal geometry appears to be particularly well suited to a first attempt at implementing this strategy.

In a preliminary way, we have tried this approach and have seen it not only enhance mathematical understanding and inquiry in students already deeply interested in and proficient at mathematics, but also to engage many students who did not see themselves as liking mathematics, by allowing them to experience mathematics more the way mathematicians do.

But preliminary trials are not enough. New decisions about curriculum should be informed by research, and we will be undertaking a systematic development of this approach and an investigation of the contribution it can make to secondary education. Some of the issues to be investigated and the possible implications for curriculum will also be discussed.

A Scenario

Imagine a group of high-school students investigating a geometric construction in which "child" triangles are attached at the corners of their "parent" triangles, each child being exactly half the size of its parent. Figures 1a through 1c show the first three stages in the development of one such construction. Figure 1d shows the seventh stage.

The students use a computer with high-speed interactive graphics tools to generate the constructions, to help them measure these constructions in various ways, to explore the effects of continuous parameter changes through

▪ Figure 1

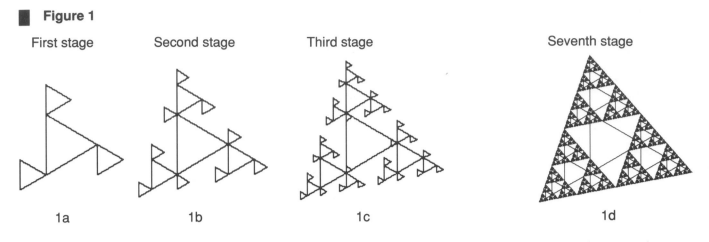

First stage   Second stage   Third stage   Seventh stage

1a   1b   1c   1d

the medium of animated sequences of figures and, in general, to experiment in ways that reflect and lead to problem posing and conjecturing and that support the development of proofs.

The teacher helps elicit the students' observations, questions, and conjectures about the frames in Figure 1. The teacher also helps the students to articulate their observations and reformulate them in ways that can be explored mathematically. Not surprisingly, students typically begin by noting attributes they have been taught to observe, and in geometry this means that the earliest observations focus on *count*, *perimeter*, and *area*. The complexity of these fractal figures, however, allows students considerable flexibility about what to count or what area they pick to measure. Here are some examples, several of which I and my colleagues have actually observed in groups of students with whom we have worked.

Posing Problems by Observing and Conjecturing

Counting Triangles. Look at the eight tiny triangles that line up along the base of the third stage. In a line parallel to them and just above them are four tiny triangles. Above *them* are another four. The number of triangles in successive rows is 8, 4, 4, 2, 4, 2, 2, 1 — all powers of two. Their sum is 27, a power of three. The powers of two in each row and a power of three for a sum can be observed in the first and second stages. Trying to predict the *sequence* in which the powers of two appear for any stage, students may try counting how many of each power appear, and notice entries from Pascal's triangle. Another intriguing and informative pattern would have remained unnoticed by students unless they had had, earlier in the curriculum, encountered a related pattern in the number of 1s in binary representations of the integers. Having begun with a geometric construction, we now find ourselves exploring number theory.

Counting Segments Along the Perimeter. If we "shrink-wrapped" the first stage, the wrapper would have nine sides. (The zeroth stage, a parent with no children, would only have three sides.) The second stage's wrapper would be more convoluted with 33 sides, and the third stage has even more wrinkles in its skin, with 93 sides. Yet the seventh stage suggests that, at the limit, the outside triangles might be growing together to give the figure a perfectly smooth triangular border: a shrink wrap with only three sides, just like the zeroth stage! What actually *is* happening? Does the sequence 3, 9, 33, 93, ... really end with a 3?! That is, do the outermost triangles actually grow together *precisely* to make a truly straight edge? Or do there remain gaps even at the limit, leaving the border porous and infinitely convoluted? Do the edge triangles even grow right over each other, overlapping at some point rather than meeting neatly?

This exploration takes some really wild twists and turns. With a very little geometry and an understanding of the sum 1/2 + 1/4 + 1/8 + ⋯, one can show that the outermost triangles do grow together precisely; i.e., that the points that look like they are going to meet *do* meet. So the border seems quite solid. But a slightly different approach, looking at the coordinates of the approaching corners, reraises the question about whether the border is porous or not, this time making that a much more sophisticated question than it originally seemed.

The Geometry of the Perimeter. Whatever the exact nature of the edge of this figure, its *appearance* is clearly a triangle similar to the root (original parent) triangle but oriented differently. What is the size ratio between the two? At what angle is the border triangle tilted? If the root were not equilateral, how would the border turn out? Similar to the original? Still equilateral? Ragged? Is it possible to build an embedded triangle fractal whose border and root are oriented the *same* way?

The Figure Considered as an Area. The pattern of light and dark regions in the seventh stage suggests a different construction: a triangle with its center fourth removed, and then the center fourths of the remaining three quarter-triangles removed, and so on, infinitely. If we had constructed the figure by successive removals of a portion of its area, how can we describe the space that remains?

Investigating the Problems

Individually or in groups of two or three, students systematically investigate some of their questions.

Exploring the Border: from Qualitative to Quantitative Argument. One group decides to examine the border of the limit for "holes" but is uncertain about how to begin. The teacher suggests that they perform experiments with various reduction rates (ratios of parent to child size) to convince themselves that sufficiently great reduction produces sparse images with borders that are certainly full of holes, while sufficiently low rates produce "crammed" pictures in which the border triangles overlap one another. Students perform the experiments and then use mathematical reasoning to check their conjecture that 2 is the exact reduction rate for which the border triangles, at the limit, just meet.

Exploring Size and Angle. A second group is concerned about the resolution of the design on the computer screen and decides to find the starting position and size that would allow them to draw the largest version they can of the stage 7 design (Figure 1d, repeated as Figure 2). This requires measuring the size of the triangular envelope of the limit figure, and accounting for its different orientation from the base triangle.

In tackling this problem, the students begin with elementary algebra and geometry and learn new mathematical ideas, including sums of a geometric series and the law of cosines.

▪ Figures 2 and 3

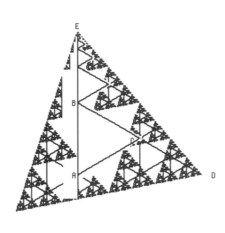

Figure 3 suggests the beginning of the analysis of this envelope. The students begin by erasing some of the distracting underbrush around the original parent triangle *ABC* (*AB*=*BC*=*CA*=1) and one series of its descendents growing along line *BE*. Then the students focus their attention on the segment *BE*. Because each "child" triangle grows to half the size of its parent, the lengths of the segments along *BE* are, in order, 1/2, 1/4, 1/8, etc. The limit of length *BE* is the infinite sum 1/2 + 1/4 + ⋯ . Using this method, the students determine the lengths *BE* = 1 and *BD* = 2. That, and the knowledge that angle *EBD*=120°, creates a comfortable and practical place to generalize the Pythagorean theorem to the Law of Cosines, and to apply it to find that the length of *ED* in triangle *EBD* is √7.

▪ Figure 4

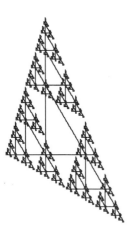

Starting with Different Root Triangles. A third group reproduces this construction using a 30°-60° triangle instead of an equilateral triangle as the root, and discovers that the border again limits to a triangle. But this triangle is neither similar *nor* equilateral! Looking at their picture (Figure 4), they conjecture that the border of this fractal is isosceles, but they ultimately disprove their conjecture using the same methods that the second group developed in its measurement project. This group's observations also spark other students to try nonequilateral root triangles.

Growing Children at Different Orientations from the Parent. Another group decides to explore the orientation of the border. They try to change its orientation with respect to the original triangle by changing the angle at which the children are attached. For their first experiment, they pick a "neat" angle, bending each child 30° to the left. The first stages of this process are shown in Figure 5a-c; the seventh stage is shown in Figure 5d.

This group's work generates many new observations and questions. Clearly the border is not a triangle. One student says it looks vaguely like an ivy leaf. The group of students who were investigating the "smoothness" of the border in Figure 1d comment that according to the way they are now looking at borders, this figure appears not to have one at all. Some students conjecture that a different reduction rate might cause the border to coalesce, while others become interested in the intermediate stages between Figures 1d and 5d, and

▪ Figure 5

First stage   Second stage   Third stage   Seventh stage

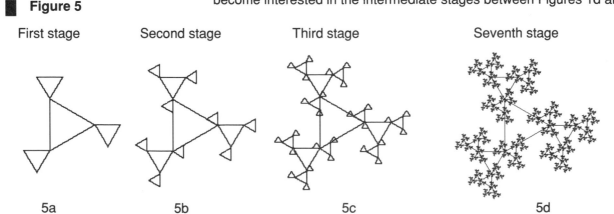

5a   5b   5c   5d

start making an animation, frame by frame, to show the effect of continuous change. (See Figure 23 for a partial example of such an animation.)

The Teacher's Role

In the vignette above, the teacher provides essential guidance: introducing software, suggesting provocative problem domains, helping students learn to observe and make conjectures, and helping as an experienced senior partner whenever student investigations require mathematical knowledge or tools that they do not yet have.

For example, in the first exploration of size and angle, although it was no more immediately apparent to the teacher than to the students how to measure the outside envelope, the teacher's involvement in brainstorming with the students about places to get a foothold on this complex design was probably essential. Together they found the measurable triangle *EBD*. When my colleagues and I worked with a group of students on a similar problem, we found we had to teach the law of cosines, a mathematical tool that the students lacked but would use quite often in measuring figures of this kind.

But the teacher's role changes dramatically. Using the metaphor of college professional ranking, the teacher is promoted from lecturer or instructor to a more creative research position: professor as mentor to a group of more autonomous students.

The students' own observations and conjectures, for the most part, fuel the investigations and determine their direction. Using well-designed software tools and appropriate metaphors, students have the opportunity to switch naturally among visual, experimental, and formal investigative styles. They are taught to make new conjectures and find new problems as they work on old ones. And students draw from conventional mathematical territories while making forays into unfamiliar contemporary domains. Rather than practicing learned computations on canned problems, these students and their teacher have the experience of *finding* new problems, exploring unknown territory, and inventing their own methods for grappling with the problems (including, of course, acquiring known tools that bear on the problem). Key motivational issues are already solved. The meaning and purpose of summing of infinite series, for example, become quite clear in a domain that consists of infinite geometrical processes.

Examples of Student Work

This sampler of fractals projects touches on a variety of mathematical topics, such as area, infinity, and measure theory. Most of the projects are based on the author's informal, exploratory teaching experiments with middle school and high school students. We believe these projects form the basis for interesting and challenging classroom explorations.

Example 1: Growing Weeds

Key mathematical concepts: scientific modeling, qualitative reasoning, visual proof. Tried with 9th grade students.

Students became interested in modeling the effect of an external force on the appearance of a weed they had constructed, to make, for example, a windswept weed or a weeping willow bowed by gravity.

An initial simplification of gravity disregards the mass of the weed and the bending within branches and looks only at the angle at which each branch veers off from the last.[4] Because the specifications for the weeds had been in

Figure 6

6a

6b

6c

"Turtle geometry,"[5] it was natural to focus on branching angles and not on changes within the sticks from which the weed was constructed. Gravity, then, alters the branching angle of branches (that do not extend straight up or down) by pulling them down some amount, but the constraint of turtle geometry asks us to cast the *downward* turning in terms of *left* or *right* turning. A further attempt to quantify turning restricts it to only *one* direction, but in "positive" or "negative" amounts.

Without considering exact values, the students now begin taking qualitative "data" by doing experiments in their heads.

initial heading of branch	effect of gravity
upright (0°)	no effect
extended straight out to the right (90°)	maximum turning to the right
straight down (180°)	no effect
extended straight out to the left (270°)	maximum turning to the left (that is, maximum negative turning to the right)

This table expresses the effect of gravity as a function of the zero-gravity heading of the branch. Students sketched this function and were intrigued to learn that their function closely resembled another function that describes the ratios of sides in a right triangle—the sine function. A first cut at the effect of gravity, *right :force * sin heading*, was all that was necessary to distinguish the "willow" (Figure 6a), with the force set to 20, from the tree as it was designed in zero gravity (Figure 6b). To experiment with antigravity, some students made force -10. The result resembled a poplar tree (Figure 6c), but students were uncertain that the branching structure was right. By tinkering with the branching structure, student found good models of poplars but *only* when they used moderately strong "antigravity responses." What does that suggest about poplars?

Despite success with this downward force, the students had a terrible struggle with an east wind. Initial attempts to account for the different direction of the force led to experiments like *right :force * sin (heading–90)*, but these didn't work. The visual effect showed them where their misconception was.

Example 2: The Sierpinski gasket

Key mathematical topics: paradox, generalization, surprise, visual reasoning, infinity.
Tried with 9th through 12th grade students in a heterogeneous mix.

Some patterns occur surprisingly often. The pattern of sparse regions in Figure 2 is highlighted in Figure 7. That is the same pattern one sees as the result of another quite different experiment with triangles (Figure 8), and is *also* the pattern of odd and even numbers in Pascal's triangle. Several students, seeing this recurring figure, asked what these three (and many other) very different generating rules have in common. Explorations of this figure, the Sierpinski gasket, lead not only to traditional mathematical content — sums of infinite series, modular arithmetic, and the like — but to novel questions that clearly intrigue students, and which are nevertheless accessible even to relative beginners. Asking how much of the whole figure's area is contained in the sparse regions is a bit like asking what proportion of Pascal's triangle is even. Grappling with the paradox that appears to claim that all of the numbers in Pascal's triangle are even can lead to insights about infinite versus finite domains. Presented intuitively at first, concepts of fractional dimension are also quite reasonably explored.

■ **Figures 7 and 8**

Example 3: The chaos game

Key mathematical concept: qualitative argument.
Targeted for 8th-10th grades.

■ **Figure 9**

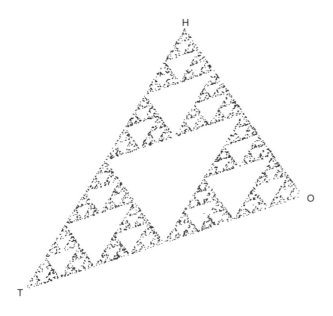

Remarkably, the very same pattern results from the following game. Picture yourself jumping around on a plane using an inked pogo stick (one that stamps a dot wherever it lands), while the following rules govern your jumps. Three points defining a triangle on the plane are labeled *H*, *T*, and *O*. Start your pogo-stick jumping anywhere: inside, outside, or on the triangle defined by the first three points. Flip a three-sided coin, one with a *H*eads side, a *T*ails side, and an *O*ther side. If the coin comes up *H*eads, jump from wherever you are to the point halfway between you and *H*. Treat *T*ails and *O*ther similarly: if the coin comes up *T*ails, jump half way from wherever you are to *T*, and if it comes up on the *O*ther side jump half way to *O*. If you simulate this game on the computer, you see essentially the same pattern, no matter how you pick the points *H*, *T*, and *O*, no matter where you start the pogo-stick, and no matter what order the "coin" lands on its three sides (Figure 9). It is possible to use purely qualitative arguments and the most elementary geometry to explain why this pattern seems to be utterly insensitive to the initial location of the pogo stick and the randomness introduced by the coin. It is also interesting to explore what aspects of this process are *not* insensitive to starting conditions!

The mathematics here is potentially very deep, yet requires only elementary geometry. Problem posing here is also very rich. The few students with whom we have tried these experiments naturally began altering the rules. What if there were four starting points? What if one jumped two-thirds of the way instead of halfway each time? What if on were to jump *away* from some of the points and toward others? What if one rotated around some or all of the points? And so on.

Example 4: Dissections of a square

Key mathematical concepts: qualitative reasoning and using invariance under transformation to think about limit behaviors, infinity.
Tried with heterogeneous class of 6th grade students.

Figure 10

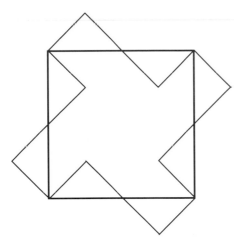

A 6th grade class investigated a transformation on a square in which a small triangular chunk was cut out of each side and attached outside the square along the remaining portion of the same side (see Figure 10). The light zigzag on each side of the square (dark) dips into the square to the right of the midpoint and out of the square to the left of the midpoint.

Area remains invariant under this transformation. Initially by hand, the students then repeated the same transformation on the resulting cross-like figure: the midpoint of each edge was marked and a triangular chunk to the right of that midpoint was removed from inside the figure and grafted back on the outside to the left of the midpoint. A sequence of two of these figures appears in Figure 11. In each case, the transformation rearranges the area but leaves it unchanged. What happens to the perimeter? Students in the 6th grade were able to generate good arguments that the perimeter would grow in an unbounded way, while the area remained constant. They compared this to another pathological curve — the snowflake. Again, they showed the perimeter was unbounded. In this case the area *also* grew. However, the students were able to generate solid qualitative arguments — and, most importantly, *cared deeply to make these arguments* — that the area was *not* unbounded, even though they could not figure out quite what it was.

Figure 11

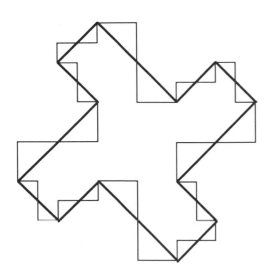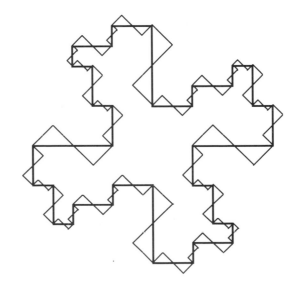

Example 5: The Koch curve

Key mathematical concepts: experimentation leading to novel discoveries and the invention of a symbolic representation to solve a problem.
Tried with one 12th grade student.

A high-school senior who was left, at first, entirely to his own devices approached the question of boundary length on the Koch curve shown in Figure 11 in what seemed initially to be a most inept and unmathematical way. Instead of analyzing how each transformation changed the length of the curve, he began by asking the computer to keep a record of the lengths of the line segments it made as it drew the curve and then add up that list. For technical reasons, he could not get the computer to add the very long lists successfully, so he tried a different approach, keeping a frequency table for each length segment the computer drew. Concentrating on a single side of the square, he listed the untransformed side as containing only one length. The first transformation (see Figure 10) had three segments of two different lengths — one long segment, and two others that were half its size. The second transformation (Figure 11a) contained nine segments of three different lengths, again only one longest segment, but this time four each of the two other sizes. He listed these data as shown in table 1:

1						untransformed
1	2					1st transformation
1	4	4				2nd transformation
1	6	12	8			3rd transformation
1	8	24	32	16		4th transformation
1	10	40	80	80	32	5th transformation

Table 1

Because this pattern was, itself, so interesting, he pondered it for a while, but except for the obvious pattern in the first and second columns and the powers of two that head each column, he could do little with it. So he returned to his original question and took the next step, multiplying each frequency listing by the length of the segment that it counted:

100.00				
70.71	70.71			
50.00	100.00	50.00		
35.36	106.07	106.07	35.36	
25.00	100.00	150.00	100.00	25.00

Table 2

The unexpected symmetry again aroused his curiosity, and he decided to look more closely at the pattern by, as he put it, "putting each row on the same scale." Dividing the entries in each row by the smallest number in the row (the length of the longest segment), he produced a table of numbers that was suddenly and stunningly familiar: Pascal's triangle!

This result led to a whole new investigation and a symbolic analysis of the geometric transformation. Influenced by the relationship between Pascal's triangle and the binomial theorem, and with some guidance from the teacher, he saw the iteration of this transformation as "raising the power" of the transformation. Representing the original untransformed segment as 1, and the first transformed set of segments (one long and two short) as ($1l+2s$), the student explored powers of this ($l+2s$) transformation. The coefficients of ($l+2s$)^n were the entries he found in his original data in Table 1.

Precisely because this student eventually took a remarkably sophisticated course in his analysis, using a symbolic system to model the geometric one with which he had begun, it is important to remember that this mathematically very rich interplay of geometric, numerical, and symbolic analyses evolved out of the failure of an initial approach that was awkward and nonanalytic.

Selecting a Domain —— Why Fractal Geometry?

Fractal geometry combines several important features that make it an exciting example of a living, changing mathematics appropriate to secondary curriculum and an ideal first test of the focus on experimental and visual mathematics and problem posing.

Attracting Students to the Intellectual Beauty of Mathematics

Fractal geometry studies complex natural and fantastic forms that are often strikingly beautiful. The very notion that complex natural behaviors can result from simple mathematical rules appears to captivate popular imagination, as evidenced by the success of Gleick's [10] *Chaos*. When a book about *mathematics* can be a national bestseller for months, its popularity testifies not merely to the author's skill, but to the provocative ideas contained in the book. Our own exploratory work with high school students also provides evidence of the inherent appeal of fractals as an avenue for mathematics learning in classrooms. We may exploit the visual beauty of fractals, their ubiquity in nature, and especially the remarkable ways they develop, support, and defy our intuitions, to attract students to the intellectual beauty of the mathematics behind the fractals.

Integrating Mathematics and Science Curricula

Fractal geometry has been recognized as a highly prized modeling tool, applicable in a wide variety of sciences.[6] The forms seen in trees, vascular systems, rivers, mineral veins, coastlines, mountains, clouds, and textures comprise just a fraction of the diversity of forms this geometry can describe. The "messy" data of economics, interference in data transmissions, heart rhythms, fluctuations in regional atmospheric temperature, the distribution of craters on the moon, and other complex forms are often modeled more closely by fractals than by traditional statistical measures using Euclidean metrics. Considering why self-similarity is predictable in certain kinds of physical processes (e.g., the formation of a mineral vein) and biological processes (e.g., the branching and forking of trees) offers intriguing topics for student research and contemplation. These broad applications in science attest to the importance of fractal geometry as a tool beyond the realm of academic mathematics and its potentially pivotal position in the curriculum as an organizing and unifying force for science and mathematics.

A Scaffold for Reasoning and a Matrix for Problem-Posing: Creating Mathematics

Fractal geometry connects closely with both traditional and contemporary mathematical domains.[7] It can enhance the most important elements of traditional mathematical learning by, for example, helping students develop a more robust concept of function. It is a natural context within which to introduce and apply many traditional high school topics (e.g., trigonometric functions, notions of limits), and it can help show the interrelationship of such topics as algebra and plane geometry which often appear distinct and isolated. At the same time, fractal geometry is an active field in 20th century mathematical research and it leads naturally toward other contemporary domains by suggesting extensions and generalizations of such familiar notions as length, area, dimension, space, and randomness. Such generalization is, in itself, an important mathematical activity, while the domains that are invoked by such generalization — measure theory, studies of information complexity, and others — may represent valuable new content or suggest directions for further investigation.

On the surface, it seems perhaps unreasonable to think of teaching to junior-high and early high-school students topics from measure theory, calculus, trigonometry, and other "advanced" mathematical territories that arise in the course of studying fractal geometry. But mathematical experimentation in a suitable environment can provide a kind of clarity and concreteness that acts as a scaffold for knowledge and reasoning and thus makes a broader range of topics accessible. The fact that fractal geometry deals in the *visual* realm of iterated geometrical processes may also contribute to its accessibility to even younger or more mathematically naive students than could handle studies of iterated *numerical* processes. Symbolic systems that are developed to represent or manipulate ideas or processes in fractal geometry can draw some of their meaning from the visual realm in which the ideas or processes reside.

But far more importantly, by creating a computer-based environment that supports students' asking of genuinely new questions and of performing actual mathematical research we exploit fractal geometry's newness and its visual/experimental nature to foster students' exploration and creation of mathematics. Beyond greatly enhancing the traditional symbolic deductive reasoning style by augmenting symbolic arguments with a semantics enriched by pictures, we expect that the full use of interactive graphics may also help students develop and use alternative modes of reasoning based centrally on visual and qualitative argument. Experimentation and visual reasoning are important alternatives to the usual style of symbolic-deductive mathematical argument. Good qualitative and visual arguments are often precursors to (and motivators for) more rigorous mathematical proofs, and are sometimes fully sufficient proofs in themselves.

Reasoning based on a visual/exploratory approach and applied to a geometry of forms more complex than circles and squares can allow precollege students to experience the *finding, posing, and articulating* of problems in a way that is typically unavailable to students until they are approaching graduate level studies. To a broad class of students, grappling with a kind of messy real-world complexity is often more attractive than acquiring techniques that "will be useful later," even though the resulting terrain is full of very hard problems that the students can solve only partially if at all.[8] With appropriate structure and guidance, the experience of *finding* problems and the spirit of *posing* problems should help students develop greater investment and satisfaction in trying to *solve* the problems while at the same time giving them important perspectives on *how* to solve the problems.[9] Further, when students experience the kind of conflict that leads them to wish to alter a definition to accommodate a new generalization of an old idea, they may no longer see mathematics as rigid, unchanging, unforgiving, and finished, but rather as a live and evolving study. Such a view is a necessary precondition both to serious mathematical research and to the flexible and creative application of mathematics in other fields.

The activities discussed above, the finding or posing of problems and the making or changing of definitions, are fundamentally acts of *creating* mathematics and are important reasoning skills in their own right. This content and pedagogical style should foster still other acts of mathematical generativity, in which students abandon the expectation that their job is to apply known methods to solve problems posed by others, and instead adopt a new stance with an autonomous drive to find and pose their own problems. This stance of "self-propelled inquiry" has at its center the interest *and knowledge* to tinker with an existing system to find out what happens when a feature is changed and to build genuinely new mathematical systems.

Helping students achieve true autonomy takes work and guidance on the part of teachers and curriculum developers. Self-propelled *guided* inquiry helps students avoid the often frustrating experience of spinning their wheels rather than propelling themselves in any recognizable direction.

Preliminary work that we have conducted with students in 7th through 12th grades and our work with high school mathematics teachers strongly suggest that fractal geometry offers a powerful, viable area for junior and senior high school students. Using a guided inquiry approach as a basis for students' explorations of fractals, we have observed a wide variety of students make significant attitudinal and stylistic changes: exploring avidly, generating and refining problems for themselves and others, and investing themselves in developing competence in intellectually challenging traditional and contemporary mathematical domains. Such interest and independent problem-posing was observed equally among boys and girls, and among both students who had generally done poorly or avoided mathematics and "A" students whose success in mathematics had depended on their competence as syntacticians and good memorizers.

Existing Intellectual Bases

To the best of our knowledge, no curriculum for grades 7 through 12 takes the approach that we propose, incorporating an important 20th century mathematical domain and a deliberate integration of problem posing and experimental mathematics. Bringing the culture and vitality of university-level mathematical research to students in the secondary classroom is an ambitious task, but it is one that draws support from the various parallel efforts of others. There is a well-developed intellectual base upon which our curriculum can draw.

Turtle-geometric strategies for generating fractals have been described as far back as Abelson and diSessa (1981) [1]. My colleagues and I have been involved for many years in developing student explorations of fractal forms using turtle geometry, and we expect that turtle geometry will remain an important introductory fractal-building tool in this project. Another source from which to draw curricular ideas and organization is Barnsley's [2] comprehensive introduction to fractal geometry, which begins with an introduction to metric spaces and focusses heavily on iterated space-mapping functions. Although this book is written at an advanced college level, it provides the kind of organization that can help identify interesting areas for investigation. With appropriate tools and metaphors, many portions of this otherwise advanced curriculum may well become accessible in a 7th through 12th grade environment. It, therefore, makes sense to explore what portions of the newly accessible material would best serve the mathematical development, in particular the development of mathematical curiosity, of secondary students. In a similar way, Feder's text [9], which deals heavily with fractals in the sciences and is again aimed at more advanced college levels, may well serve as a resource, organizer, and guide for appropriately selected work at earlier levels.[10]

Finally, there are related ongoing curriculum research projects with which rich connections can be made. Among others, both Boston University (see Devaney [7]) and BBN Laboratories are funded by the NSF to bring the mathematics of chaos and nonlinear dynamics to high-school students.[11]

Three Sources of Fractal Geometry's Impact on the Curriculum

Visual Problem Posing and Experimental Mathematics

When a delicate lacework of triangles (Figure 12) or a believable weed (Figure 13) begins to emerge as a computer executes a simple and easily understood rule, onlookers are almost always transfixed. The result is beautiful, the process is entrancing, and, as evidenced in our preliminary work with students, the mathematics, too, is captivating. Students eagerly begin to suggest alternative rules, needing no coaxing from teachers.

■ **Figures 12 and 13**

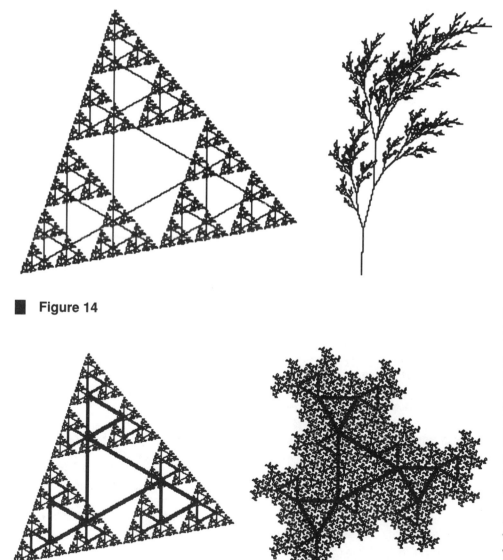

■ **Figure 14**

14a   14b

For example, the triangle lacework was generated by a rule that affixes "child" triangles at the corners of "parents" (see highlighting in Figure 14a). Students readily ask: What if we changed the angle at which these smaller triangles were affixed? Instead of attaching them "straight" at the corners of the "parent" (as in Figures 12 and 14a), we might bend them 30° to the left (14b).

Or what if we began as in Figures 12 and 14a, but allowed "children" to sprout only from a single corner of the parent triangle? One such case is shown in Figure 15. This geometrical process seems to converge on a single point. Where is that point in relation to the frame around the picture? The current parent/child size ratio is 2:1. Is there a different ratio for which the process would converge at the horizontal midpoint of the frame?

Similarly, can we make the weed's stalk bend, as if weighed down by gravity (Figure 16a)? Or what if it resists "gravity" (Figure 16c)?

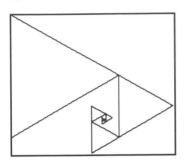

Figure 15

Figure 16: *Three trees with identical branching structure, differing only in their response to simulated gravity*

The fact that the weeds in Figures 16a, b, and c, have *identical* branching rules and differ only in their response to "gravity" suggests some interesting questions about the role that the plant's building material plays in the visual expression of its form. Figure 16a looks more flexible and is a more reasonable model of a grassy plant than Figure 16c, which might be taken for an asymmetric poplar rather than a grass.

Such problem posing generally begins at the visual level as a kind of tinkering with the image. In our preliminary work, the students showed intense interest in the graphic experiments and, especially in the earliest stages, often performed the experiments out of a kind of naive curiosity without having formulated much in the way of expectations about the outcome. As they became more familiar with the results, and especially as their attention was drawn to particular features of these results, their approach changed and they tried to achieve specific effects. The first such attempts represent crude tests of partially articulated conjectures, but they also indicate the emergence of self-motivated problem posing in the students. And right from the start, they suggest questions about science as well as mathematics

16a

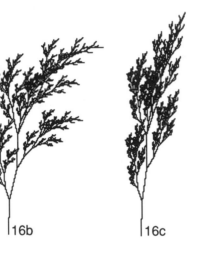

16b   16c

This attitude of experimentation — the "what if" and "what if not" frame of mind — is an essential ingredient in mathematical creativity and research. It generates problems to solve, fosters conjecturing, and often leads students to increasingly more sophisticated questions and generalizations, as well as an interest in verifying those generalizations. The earliest development of mathematical questions may arise out of very practical problems like how best to fit a figure on the computer screen. To get the greatest detail, we might want to start with the largest figure we can create. This involves understanding, in at least a crude way, how the figure's limit size behaves. How tall, for example, is the entire weed compared to the first segment of its stem? Or, how large can we make the central triangle in Figure 14a and still have the entire figure fit on the screen? And, where should that triangle be placed? Questions of this kind lead naturally into some conventional mathematical territory: limit behaviors, series, some geometry, and trigonometry.

Rethinking Old Concepts

Topics from contemporary mathematics evolve from a different kind of observation. Students spontaneously wonder at how the enormous visual difference between Figures 14a and 14b results from the seemingly benign change in the angle at which the offspring triangles are attached to the parent. Figure 14a is very geometrical in appearance: it has the outline of a triangle, with many internal empty regions, also triangular. Figure 14b (ignoring the added highlights that show how it is generated) is quite uniform throughout and is bounded by a curve that seems to be a stylized derivative of some natural object, perhaps a leaf or an ocean wave. Mathematizing these observations is not so spontaneous, but is, in our experience, easily encouraged. The observations themselves are so compelling that as the students learn a new style of questioning they seem to take it on quite readily, especially since the questions lead to such intriguing and thought-provoking answers.

For example, how much of Figure 14a is taken up by the nearly empty regions? The largest of these regions is 1/4 of the entire figure. Each of the three remaining quarters contains an empty region that is 1/4 of *it*. Each of the nine remaining regions contains an empty region that is 1/4 of *it*, and so on. Summing this series makes it appear that *all* of the figure is empty, a conclusion that, on the face, seems quite surprising.

Furthermore, even though the combined empty regions of Figure 14a seem to contain all the *area* of the enveloping external triangle, they clearly do not include every *point* inside that envelope. By contrast, Figure 14b seems quite full. *How* full? Does the limit curve cover *all* of the region internal to the "shrink-wrap" envelope that surrounds it? And if so, what would that mean? And what about the "length" of the envelope? Figure 14a appears, at each successive step in its development, to be smoother and smoother around its perimeter and to approximate more and more closely an equilateral triangle of side $\sqrt{7}$ more and more closely.[12] If we imagine the limit fractal (infinite iteration) and examine any portion of it under a microscope, it appears just the same. Consequently the shrink-wrapped edge appears just as smooth. The fractal in Figure 14b exhibits the same kind of self-similarity, but maintains the infinite crinkliness of its shrink-wrap edge at all scales. The "length" of that edge appears to be infinite, though perhaps "length," at least as we commonly think about it, is the wrong measure for such a curve. Our concepts of length, area, volume, and dimension must expand as we consider such objects. Some figures have measures that can be computed precisely, but cannot be determined by any measuring unit. Other figures seem to defy measure altogether. *Measure*, itself, becomes an object of interest.

With experience, encouragement, and instruction, participants in this kind of exploration develop a sense that *mathematical* verification requires *proof*, but that the generative process may draw importantly from formal, informal, and experimental methods.

Enhancing Reasoning with Visual and Qualitative Thinking

To provide even one clear example of what we mean by visual and qualitative reasoning requires a somewhat lengthy discussion (and a single example can never capture the breadth of application of this kind of reasoning), but it seems essential to illustrate the power that visual and qualitative thinking can bring to a problem that, at its outset, seems out of reach to a student with only typical high-school level mathematical tools.

> **Problem:** Show that there may be infinitely many points for which Newton's method does not converge, and that the set of these points has the kind of self-similarity (or quasi-self-similarity) that characterizes fractals.

Besides giving an opportunity to illustrate a reasoning style, this problem also exemplifies a bridge between fractal geometry and traditional mathematics. For brevity, the problem is posed here without definitions of Newton's method, explanations of convergence, etc. But that also serves to highlight another point we are making: the difficulty of a problem is in part an artifact of the context in which it is set and the tools that are suggested. To a student, this problem is more aptly posed as an iterated geometrical process of drawing tangents, locating new ones, and watching until the process seems to stop going anywhere.

Perhaps more important than *solving* this problem (visually or by any other means), students who have the opportunity to experiment with the behavior of this iterated geometrical function and its convergence rate might reasonably be led to *pose* the problem as a result of their own observations. The software tools we will build will facilitate such experimentation.

■ **Figures 17, 18 and 19**

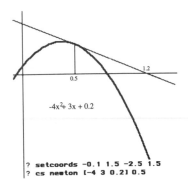 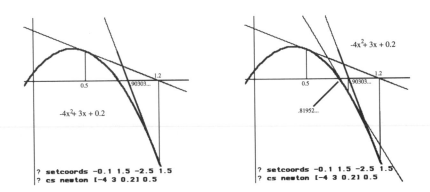

■ **Figure 20**

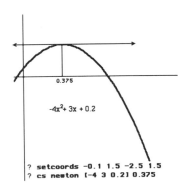

■ **Figure 21**

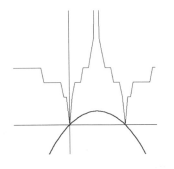

The following is an example of the kind of paths a student might take in exploring this problem.

Observe the behavior. Apply Newton's method to some function. In Figures 17-19 it is applied to $-4x^2 + 3x + 0.2$, starting at $x_0 = 0.5$. The tangent at $(x_0, f(x_0))$ intersects the x axis at $x_1 = 1.2$ (Figure 17).[13] The second tangent is drawn at $(x_1, f(x_1))$ and intersects the x axis at $x_2 = 0.90303...$ (Figure 18). The third tangent is drawn at $(x_2, f(x_2))$ and intersects the x axis at $x_3 = 0.81952...$ (Figure 19).

Notice a pattern and a problem. In the experiment from which Figures 17-19 were drawn, the tangents converged. After the third iteration (Figure 19) each new tangent seemed to be drawn on top of the previous one. However, if we had started the process in a different place, the process would not converge. For example, at a local extremum the tangent is horizontal and therefore will never intersect the x axis (Figure 20). The process stops abruptly. If we view the process geometrically, we conjecture (correctly, but *only* for this example) that any deviation from this local maximum will converge, and that the *rate* of convergence should be slow if we are too close to the local maximum.

Graph the rate of convergence starting at different points. The smooth curve in Figure 21 is $f(x)$. The step-graph above it shows the number of iterations of Newton's method before tangents appear to lie one on top of the next (at this scale). Of course, when one starts at a root, it takes *no* iterations to get close, so the rate graph dips to zero. The rest of this graph also fits our intuition. Near the local maximum, the number of iterations before convergence is high. Between instant success and guaranteed failure are gradations.

But other curves show unanticipated complexity. For example the graph in Figure 22 seems to behave as we expect to the left of the axis, but has unexpected features to the right. Tangents at these features clarify what is going on. A suitably picked tangent near the right side's unexpected dip leads directly to the root on the left-hand side — instant success. But the tangent from the right side's unexpected peak leads to the left side's local minimum — instant failure. Tangents from unexpected features on the right lead directly to expected features on the left. Thus, the left-side terrain is reflected into the right side.

Argue for "symmetry." Having understood that the behavior on the right includes a reflected copy of the behavior on the left, one might reasonably ask why we do not see similar complexity on the left! Why, after all, should there be an asymmetry? In fact, there isn't. Increasing the resolution on the left-hand peak shows that it, too, has "unexpected" features that reflect behavior on the right. This recursive situation (each side contains copies of the other side, including the other side's "unexpected" features) describes either a cyclic behavior or an infinitely embedded structure with the kind of quasi-self-similarity that we sought to demonstrate. Students can create enlarged graphs at greater resolution to show the predicted embeddings.

Figure 22

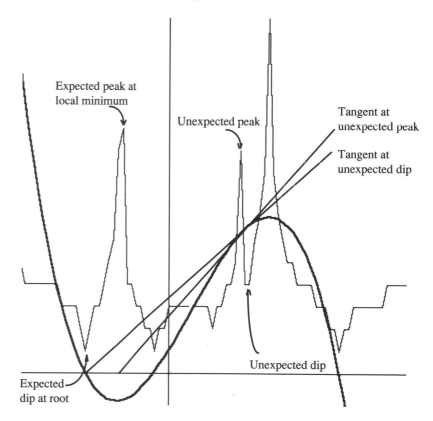

Expected peak at local minimum

Unexpected peak

Tangent at unexpected peak

Tangent at unexpected dip

Unexpected dip

Expected dip at root

Nuts and Bolts: Crafting a Curriculum that Fosters Exploration

To develop this curriculum, we must identify accessible and appropriate topics for 7th through 12th grade students, including both fractal geometry's unique contributions and those traditional mathematical topics that may arise from the study of fractal forms. The curriculum must also establish connections with biology, geology, physics, and other sciences, especially looking for those places where studies of fractals can lead to important insights and a unified view of mathematics and science. This exploration will guide the development of educational materials and software tools.

Appropriate Structure

Because the goal is a kind of intellectual autonomy for students, we must straddle a thin line between prescribing every student move and leaving students so unguided that they flounder too early and, for that reason, fear to venture out alone. We must therefore consider what kind of structure is most helpful. Having identified an initial set of topics, we envision a set of starting ideas and materials, both computer-based and off-line, with two pedagogical goals in mind: to guide initial exploration (using a blend of explanatory background and fully prescribed investigations); and to provide a supportive structure aimed at helping students develop their ability to explore autonomously in a more open-ended and self-initiated way.

Here are two examples that suggest the prescriptive structure of the investigations.

Etude 1: Equivalent and nonequivalent rules for building a triangle

Here is a simple rule for making a triangle:

(1) *repeat 3 [forward :s right 120]*
> Repeat three times a process that moves the drawing stylus forward a distance of :s in the direction it is heading, and then turns the heading 120° to the right.

The commands in the rule can be abbreviated this way:

(1a) *repeat 3 [fd :s rt 120]*

Create this procedure that makes use of rule 1a.

```
to tri :s
   repeat 3 [fd :s  rt 120]
end
```

Now experiment with this procedure by typing commands like *tri 100* or *tri 22.5*.

We shall call two rules "equivalent" if they produce the same figure with all inputs. Three of the following four rules are equivalent to rule (1). Decide which are and which is not. Explain why the equivalent ones are equivalent and in what way the nonequivalent one draws a different picture. Modify your procedure and experiment with these rules, if you like.

(2) *repeat 3 [fd :s/2 fd :s/2 rt 120]*
(3) *repeat 3 [fd :s/2 rt 120 fd :s/2]*
(4) *repeat 3 [fd :s lt 30 rt 150]*
(5) *repeat 3 [fd :s/2 lt 30 rt 30 fd :s/2 rt 120]*

Etude 2: Writing rules for self-embedded figures

Here is a similar procedure with a single embellishment: an input that, if zero, causes the procedure to stop immediately without drawing a triangle.

(The reason for making this seemingly useless embellishment will become clear later.)

```
to tri :c :s
  if :c = 0 [stop]
    repeat 3 [fd :s  rt 120]
end
```

Create and try this procedure to be certain you understand the function of the new parts.

Now consider the following more significant embellishment. Somewhere, en route to drawing the triangle, pause, make another similar but smaller triangle, and then continue. For example, after drawing each side but before turning to draw the next side, you might draw a half-size triangle. That instruction would look like this:

(6) *repeat 3 [fd :s* **(tri :c-1 :s/2)** *rt 120]*

This rule is based on rule (1) above. The half-size triangle (its instruction is shown in boldface) is drawn after drawing the side of the current triangle but before turning to draw the next side.

The procedure to use this instruction would look like this:

```
to tri :c :s
  if :c = 0 [stop]
    repeat 3 [fd :s  (tri :c-1 :s/2)  rt 120]
end
```

and would be executed like this

tri 1 100 or *tri 2 100* or *tri 3 100* or *tri 2 200*

Run the procedure using values of :c $\leq$ 6 and any values for :s until you can explain the effects of these parameters. How does the "seemingly useless embellishment" work?

Exploration is also guided, but is much less prescriptive. Here is an example of a relatively early (and therefore still somewhat prescriptive) exploration.

Exploration 1: Alternative rules for self-embedded triangles

Here are several other interesting variations:

(7) *repeat 3 [fd :s rt 120* **(tri :c-1 :s/2)***]*
        draw the half-size triangle after turning
(8) *repeat 3 [fd :s rt 120* **(tri :c-1 :s/3)***]*
        draw a third-size triangle after turning
(9) *repeat 3 [fd :s/2* **(tri :c-1 :s/2)** *fd :s/2 rt 120]*
        start midway along a side
(10) *repeat 3 [fd :s lt 30* **(tri :c-1 :s/sqrt 3)** *rt 150]*
        from rule (4)

Try each of these. Invent some new rules that embed triangles within triangles in a new way.

More advanced explorations arise partly out of the results of students' work and are consequently even more open-ended than these. The most advanced explorations are, in effect, research projects similar in character to the work described in the classroom scenario earlier in this paper.

Different Organization from Traditional Curricula

Early experience supporting later formal study. Certain mathematical objects, such as the cosine function, would be encountered much earlier than they currently are. Initial contacts would be casual (even superficial), but the practical experience gained with these objects would make them quite homey and familiar by the time their properties are studied formally (e.g., in a course focussing attention on trigonometric functions). This, in effect, stands the conventional curriculum on its head: interesting and "difficult" problems come first, but are approached in visual, concrete, informal, and intuitive fashions, with formal tools to be acquired as they are needed. The first introduction of these tools would occur in a natural context, leaving to later courses the organization of this knowledge into formal systems and the filling-in of the inevitable gaps.

Variety as an antidote to narrow concepts. Certain important mathematical constructs would be encountered in a far deeper and more varied way than they typically are. For example, the concept of a *function*, in conventional curricula, is dealt with too superficially and students may see little other than a mapping from $\Re$ to $\Re$. Fractal geometry introduces students not only to these number-to-number functions, but also to geometrical functions that map one set of points, curves, or surfaces to another (Figure 10 is an example of $f(\text{——}) = \diagdown\diagup$), and various hybrids that map between numbers and geometrical objects. Variety, and the ability to build executable versions of these functions on the computer, should help students develop a more robust and flexible concept of function. It is also possible (as suggested in Goldenberg, 1988 [11]) that the multiplicity of images may, without considerable care, further confuse the notion of function. Each of several other important mathematical concepts — variable, measure, scale, infinity, local and global limiting behavior, iterated processes, the differential — may be strengthened by perspective or weakened by the added complexity of the new experiences. It seems reasonable to expect a benefit, but one must remain alert for misconceptions.

In fact, *both* the early experiences and the variety are likely to affect the misconceptions we see, reducing some and giving rise to others. Misconceptions can be rich grounds for expanding and clarifying mathematical notions. However, to reap the benefits, we must watch the *evolution* of student understanding and misunderstanding, and carefully design curricula that help teachers guide and support student explorations that, in turn, help students recognize and confront their misconceptions. Such close observation is supported ideally by guided inquiry and a small-group approach.

Introduction to other contemporary mathematics. We conjecture that the study of iterated geometrical functions is a good way to lay the foundation for the study of iterated numerical functions. In that way, fractal geometry, parts of which appear to be appropriate even for junior high students, may constitute a logical and exciting introduction to chaotic dynamical systems, another powerful unifier of science and mathematics curricula.

It is important to explore these issues throughout the full 7-12 grade span, in order to investigate how fractals could act as a potential steering influence on the precollege curriculum both in mathematics (a role that calculus plays in most current curricula) and science (e.g., through a study of dynamical systems). The earlier half of this age range is particularly interesting as it sheds light on possible transformations of the overall curriculum. Moreover, given educators' and parents' generally looser expectations and requirements of junior high mathematics, these grades offer greater curricular flexibility, and potentially greater impact on students' sustained engagement in mathematics.

Software Tools

What tools are available influences the structure of the curriculum and the course of a student's explorations. In designing these tools, it is important to keep the technological component simple enough so as not to distract students from the mathematics. As students need broad freedom to define their own geometric (or other) functions, they need the kind of freedom that programming permits. The sample etudes and explorations above show how student explorations may be supported by a programming language, and even give a sense of how a total novice might be introduced to nearly all of the programming knowledge needed for defining one's own functions. However, there are other essential tasks for which programming would be a distraction. These include measurement of an image; animation of a sequence of images; copying and transforming (e.g., rescaling, rotating, and translating) portions of an image to compare it with other portions of an image (for such purposes as measurement or exploration of self-similarity); changing how lines are drawn to highlight various features of an image (e.g., to show only its endpoints or midpoint), and so on.

Animation seems especially important. Figures 1 and 5 show how apparently benign changes in a single parameter can have extraordinary consequences. A sequence of still images (Figure 23) does not allow one to track the motion of various regions of the figure as one can when the images are presented at animation speed with parameters altered in a continuous fashion. Such animations suggest whole new lines of inquiry that do not arise as readily from viewing the still images.

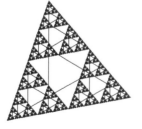

Figure 23: Four "stills" from a sequence showing the effect of angle change from 0° to 30°

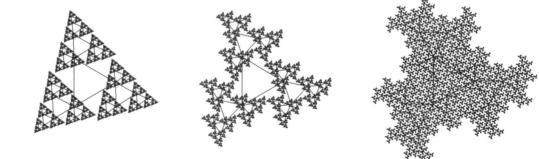

Print Materials

Software tools must be accompanied by a set of student materials, including projects and problems that lend themselves to rich explorations, sample scenarios including possible paths of inquiry, and interesting issues and further problems for student investigation. To be practical, these must be augmented with commentary and courses of instruction designed to give teachers both background and strategies that will enable them to make effective use of the student materials. In particular, attention must be focussed on helping teachers give appropriate structure and guidance to students while permitting the level of self-propelled inquiry that is so much a part of our motivation to do this work.

We will explore various formats and strategies for designing materials that best capture a guided self-propelled inquiry approach. One approach we have employed successfully (e.g., in EDC's *Visualizing Algebra* series, Harvey, Schwartz, Yerushalmy [14]) uses selected problems and teacher commentary to suggest paradigms for problem setting and analysis. This approach would include a set of paradigmatic questions and problems in the fractal geometry realm; a schema for modifying them; suggestions for helping students stretch or extend their knowledge, observational skills, or research strategies; and several examples of these in use.

Another approach (e.g., Goldenberg and Feurzeig's *Exploring Language with Logo* [12]) interweaves information, commentary, speculation, and two kinds of activities mentioned earlier: "Etudes" (a blend of explanatory background and fully prescribed investigations) and "Explorations" (suggestive but generally open-ended projects that may be attacked in more than one way). Etudes build skills that the students need in order to pursue the Explorations independently. The interweaving of these ideas may take a form similar to that employed in Goldenberg and Feurzeig [12], in which the focal subject matter (in that case, linguistics), computer-based explorations, and observations/speculations about other fields (psychology, artificial intelligence, art, science, etc.) were shown to be mutually informing. The object of this kind of strategy is to engage the teacher in the research along with the student, and to reduce the perception of mathematics as isolated from other disciplines by drawing explicit connections among science, philosophy of science, esthetics, and several branches of mathematics.

Investigating the Issues Raised by Such a Curriculum

Both the content of fractal geometry (and the topics it may give rise to) and the methods of experimental mathematics are so unexplored that a first task is to develop the approach and demonstrate its feasibility in 7th-12th grades.

However, it is essential to understand the *effect* that these contents and methods have on students' mathematical interest, generativity, and learning. We must, therefore, examine the factors that are responsible for the effects we see, generating knowledge about process as well as outcome. Of particular importance is the nature of the development of problem posing — an understanding of how problem posing evolves and a characterization of the stages in that evolution. Both "laboratory" methods and "naturalistic" measures make sense in such a study. Laboratory measures include both "metamathematical" problems such as "make up a hard math problem," or "make up an interesting math problem," or " what makes that problem hard (or interesting)?" and direct observations of student approaches to particular mathematical problems that we pose. More naturalistic measures include close attention to changes in the number and nature of the problems that students pose spontaneously.

Toward a Radical Restructuring of the Mathematics Curriculum

Seen conservatively, fractal geometry is a credible alternative to traditional precalculus courses. It ties together all prior mathematics, drawing immediately from algebra and geometry and naturally introducing elements of trigonometry and many key elements of conventional precalculus and calculus courses. A more radical vision introduces fractal geometry *early* in the curriculum, perhaps during the junior high years, in a way that suggests a fundamental restructuring of the precollege mathematics curriculum, with implications for science as well.

It is this second vision that interests us. Fractal geometry can offer an intellectually stimulating area for students, and can have a permanent and pivotal role in the secondary school curriculum. We believe it has richness and importance not only as a mathematical domain in its own right, but also as a context within which other traditional and contemporary mathematical studies may arise. In addition, fractal geometry can serve as a unifying thread between mathematics and science.

- Students' early exploration of fractals can introduce tools from much of the conventional high school mathematics curriculum, influencing the nature of later courses, while students' awareness of fractals as a modeling tool throughout the natural sciences suggests a unifying role that mathematics can play in the science curriculum.

- Fractal geometry can be seen as a powerful test case of how the 7-12 secondary school mathematics curriculum could be radically restructured by a pedagogical approach using guided inquiry, challenging and visually provocative problem domains, experimental mathematics aided by powerful high-speed interactive computer graphic tools, and an emphasis on developing students' exploratory spirit in mathematics.

The essential task of this study will be a demonstration that, given appropriate technology and curriculum, it is possible to attract a broad class of students to mathematics and mathematical ways of thinking, to foster in them self-propelled inquiry and mathematical generativity, and to teach them a body of important ideas drawn from a contemporary mathematical domain and the ways these ideas influence scientific thinking.

The primary product we seek to deliver is *knowledge* upon which we and others can base curriculum design — knowledge about a particular mathematical terrain, fractal geometry, and about a method of approaching it, visual/experimental mathematics, and, most of all, knowledge about their impact on student learning and their potential impact on the overall curriculum.

References

[1] Abelson, Harold and Andrea A. diSessa. *Turtle Geometry: The Computer as a Medium for Exploring Mathematics*. The MIT Press, Cambridge, MA, 1981.

[2] Barnsley, Michael. *Fractals Everywhere*. Academic Press, San Diego, 1988.

[3] Brown, Stephen I. and Marion I. Walter. *The Art of Problem Posing*. Lawrence Erlbaum, Hillsdale, NJ, 1983.

[4] Clark, Nigel N. "Fractal harmonics and rugged materials." *Nature*, 319(February 1986), 625.

[5] COMAP. *For All Practical Purposes*. Freeman, New York, 1988.

[6] Cuoco, A. "Visualizing the *p*-adic integers." Woburn High School, Woburn, MA. (in preparation).

[7] Devaney, Robert L. *Chaos, Fractals, and Dynamics: Computer Experiments in Mathematics.* Addison-Wesley, Redwood City, CA, 1989.

[8] Douady, R. "The interplay between the different settings, tool-object dialectic in the extension of mathematical ability." *Proceedings of the 9th International Conference for the Psychology of Mathematics Education*, State University of Utrecht, The Netherlands, Vol. II. 1985.

[9] Feder, Jens. *Fractals.* Plenum Press, New York, 1988.

[10] Gleick, James. *Chaos: Making a New Science.* Viking Penguin, New York, 1987.

[11] Goldenberg, E. Paul. "Mathematics, Metaphors, and Human Factors." *J. Mathematical Behavior, 7*(September, 1988a), 135-173.

[12] Goldenberg, E. Paul, and Wallace Feurzeig. *Exploring Language with Logo.* MIT Press, 1987.

[13] Harrison, Jenny. "Continued Fractals and the Seifert Conjecture." *Bulletin of the American Mathematical Society*, 13 (October 1985), 147-153.

[14] Harvey, Wayne, Judah L. Schwartz, and Michal Yerushalmy. *Visualizing Algebra.* Sunburst Communications, Pleasantville, NY, 1988.

[15] Heppenheimer, T. A. "Mathematicians at the receiving end." *Mosaic*, 16(1985), 37-47.

[16] Linn, Marcia C. and Petersen, Anne C. "A Meta-analysis of Gender Differences in Spatial Ability: Implications for Mathematics and Science Acheivement." In *The Psychology of Gender: Advances through Meta-analysis*. The Johns Hopkins University Press Baltimore, MD, 1988.

[17] Lovejoy, S., D. Schertzer, and P. Ladoy. "Fractal characterization of inhomogeneous geophysical measuring networks." *Nature*, 319(January 1986), 43-44.

[18] Maddox, John. "Gentle warming on fractal fashions." *Nature, 322*(July 1986), 303.

[19] Mandelbrot, Benoit B. *The Fractal Geometry of Nature.* W. H. Freeman and Company, New York, 1977, rev. ed. 1983.

[20] Moses, Barbara, Elizabeth Bjork, and E. Paul Goldenberg. "Beyond problem solving: problem posing with technology." To appear in *NCTM Yearbook, 1990.*

[21] Nittman, Johann, Gerard Daccord, and H. Eugene Stanley. "Fractal growth of viscous fingers: quantitative characterization of a fluid instability phenomenon." *Nature, 314*(March 1985), 141-144.

[22] Peitgen, Heinz-Otto and Deitmar Saupe, eds. *The Science of Fractal Images.* Springer-Verlag, New York, 1988.

[23] Peitgen, Heinz-Otto and P. H. Richter, eds. *The Beauty of Fractals: Images of complex dynamical systems*. Springer-Verlag, New York, 1986.

[24] Peitgen, H.-O. and H. Jürgens. *Fractals for the Classroom: A Short Introduction to Fractal Geometry*. Springer-Verlag, New York, 1989.

[25] Peterson, Ivars. *The Mathematical Tourist: Snapshots of Modern Mathematics.* W. H. Freeman and Company, New York, 1988.

[26] Peterson, Ivars. "Packing It In: Fractals play an important role in image compression." *Science News*, 131(May 2, 1987), 283-285.

[27] Sander, Leonard M. "Fractal growth processes." *Nature, 322*(August 1986), 789-793.

[28] Sander, Leonard M. "Viscous fingers and fractal growth." *Nature, 314*(April 1985), 405-406.

[29] Schwartz, Judah L., and Michal Yerushalmy. "The *Geometric Supposer*: using microcomputers to restore invention to the learning of mathematics." In *Contributors to thinking*. E. Perkins, J. Lochhead, and P. Butler, eds. Lawrence Erlbaum, Hillsdale, NJ (in press).

[30] Termonia, Yves and Paul Meakin. "Formation of fractal cracks in a kinetic fracture model." *Nature*, 320 (April 1986), 429-431.

[31] Turcotte, D. L., R. F. Smalley Jr. and Sara A. Solla. "Collapse of loaded fractal trees." *Nature*, 313(February 1985), 671-672.

[32] Yerushalmy, Michal and Daniel Chazan. "Effective problem posing in an inquiry environment: a case study using the *Geometric Supposer*." *Proceedings of the*

Eleventh International Conference on Psychology of Mathematics Education, vol. 2., pp. 53-60. Eds. Bergeron, Herscovics, Kieran. Montreal, 1987.

[33] Zorpette, Glenn. "Fractals: not just another pretty picture." *IEEE Spectrum*, (October 1988), 29-31.

Acknowledgements

This paper was originally published in *The Journal of Mathematical Behavior*, 8(August 1989), 169-204. It is reprinted here with permission of Ablex Publishing Corp. and the author. All of the illustrations for this paper were produced in Object Logo 2.0 for the Macintosh. EDC is currently developing tools that ease the production, measurement, and animation of these iterated geometrical objects.

Adaptation for this volume was supported in part by National Science Foundation grant MDR-8954647.

Notes

1   The Geometry Supercomputer project is one example of technology enhancing mathematical research, or see Heppenheimer [15].

2   Philip G. Lewis, personal communication

3   See, for example, Goldenberg [11], Linn and Petersen [16], Schwartz and Yerushalmy [29], Yerushalmy and Chazan [32].

4   Of course, disregarding mass makes any discussion of *gravity* meaningless, but the qualitative statements about the visual nature of gravity's effect remain reasonable, and the visual result is quite adequate.

5   All of the algorithms developed or used by students were specified in the Logo computer language. (All figures in this paper were created in Paradigm Software's *ObjectLogo* for the Apple Macintosh, but any Logo on a machine with good graphic resolution will do.) The algorithm that grew this tree is given below. The choice of Turtle Geometry's local coordinate system over a global coordinate system such as the Cartesian plane vastly simplifies the description of these geometrical objects. (The exception is in constructions such as the one represented in Example 3, Figure 9, which make explicit use of transformations on a global coordinate system. The ease with which vector and matrix operations are implemented in Logo and the fact that *ObjectLogo* provides complex arithmetic make it an ideal choice for both kinds of geometric constructions.).

```
to tree :c :s
if :c = 0 [stop]
local "root make "root list heading pos
  rt 0 grav fd :s*.2
    lt 12 tree :c-1 :s*.66 rt 12
  rt 9 grav fd :s*.1
    rt 15 tree :c-1 :s*.56 lt 15
  lt 0 grav fd :s*.1
    rt 11 tree :c-1 :s*.56 lt 11
  lt 10 grav fd :s*.2
    lt 14 tree :c-1 :s*.462 rt 14
  rt 9 grav fd :s*.2
    rt 8 tree :c-1 :s*.396 lt 8
  lt 6 grav fd :s*.2
    lt 12 tree :c-1 :s*.352 rt 12
pu seth first :root setpos last :root pd
end

to grav
rt :force * sin heading
end
```

```
cs   make "force 20   lt 3   tree 5 250
cs   make "force 0    lt 3   tree 5 250
cs   make "force -10  lt 3   tree 5 250
```

6   Perusal of just a year or so of *Nature* turns up many citations. See, e.g., Clark [4], Lovejoy, et al. [17], Maddox [18], Nittman, et al. [21], Sander [27][28], Termonia and Meakin [30], Turcotte, et al. [31]. The title of Zorpette's [35] article "Fractals: not just another pretty picture" captures the tension between the visual beauty that motivates much popular interest, and the serious importance of this mathematics.

7   E.g., Harrison [13] shows how insights from fractal geometry can shed light on a problem in number theory. Fractals also provide a helpful geometric interpretation of p-adic integers (Cuoco [6]).

8   The notion that complex problems, even if they are only partially soluble by the student, can be more interesting than procedures that a student can fully master is not a new one. COMAP's *For All Practical Purposes* [5] is an example of a mathematics text that uses this assumption as its motivation. Unfortunately, in many problem domains there is a tradeoff between interesting complexity and the ease with which students may pose problems and discover or invent even partial methods for the investigation of those problems. Thus, in the COMAP book, while one can see clearly how mathematical tools deal successfully with some enormously complex and clearly important problems, there is little sense, despite the very interesting historical vignettes in the book, that mathematics is inventable. Although the applications are interesting, students may still be left with the common impression that all problems are already staring one in the face, so to speak, and there is no need (or room) to find or pose new ones, and that one *learns* mathematical tools only from books and experts, and cannot invent them on one's own.

9   See, e.g., Moses, Bjork and Goldenberg [20]; Brown and Walter [3]; or Yerushalmy and Chazan [32].

10 The number of technical books and popularizations is growing so rapidly that the bibliography of directly relevant works of which we are already aware (a small portion of which augments our list of references at the end of this document) will increase even before this article is published.

11 Appropriate contact people are W. Feurzeig and P. Horwitz at BBN, and R. Devaney at BU.

12 A brief account of this computation and the way in which it is encountered by students can be found above in the discussion of Figure 3.

13 These tools (and the presentation to the student) do not prerequire an understanding of the $(x_0, f(x_0))$ notation, but we teach that notation.

The Visualization Environment for Mathematics Education

Steve Cunningham

This article considers three aspects of computer-based visual approaches to mathematics education. These are the state of computing technology for visualization, the role of visualization in mathematics, and barriers to visualization in mathematics education. Throughout the article, the term "visualization" will usually mean a computer-based approach, although much of the discussion also applies to visualization without computing.

"Scientific visualization" is a common current term for the use of computer graphics technology to support research in the sciences, as described in the National Science Foundation report, "Visualization in Scientific Computing" [11]. Thus visualization is not computer graphics, but it uses the tools, techniques, and devices of computer graphics to achieve its results. Scientific visualization has become something of a buzzword, but it deserves serious attention since it has become a way to discuss the use of visual approaches to understanding scientific work.

Visualization has achieved major successes in helping research scientists understand and present their results, and industrial and academic researchers are spending large sums of money to provide advanced visualization tools for science. It also offers great potential for scientific and mathematical education. This article will sketch both the technical and cultural sides of visualization with emphasis on mathematics education.

The technical side of visualization for education comprises the computing systems available for visual work in science mathematics education. It considers the computing capabilities needed for educational work and indicates to what extent these are available or under development. It finds difficulty with educational software development tools and suggests that advances in software technology may make it less difficult to develop educational software.

The cultural aspect of visualization focuses on the teaching and learning impact of visual education in mathematics. It considers properties of effective educational visualization software, curricular questions in using visualization as a major part of mathematics education, and the academic environment's effects on such curricular changes. It finds some gaps in our knowledge of teaching and learning where research is needed.

The Technology of Visualization

A wide range of technology is used for scientific visualization, from million-dollar "graphics supercomputers" to four-figure personal computers. However, we cannot afford many computers with visualization capabilities for education, so we must usually use the least expensive machine in a product line. This makes it difficult for developers to base software on anything but lowest-denominator machines. For example, almost all educational software now

Steve Cunningham is Professor of Computer Science at California State University Stanislaus. He has written widely on issues in computer graphics for education and is actively involved in both computer science and computer graphics education.

available is written for the Macintosh Plus or the IBM PC and compatibles with CGA or EGA graphics.

What does mathematics education need from computers? Our systems must perform computations quickly and return results to the students via good graphics. They must support easy and flexible user interaction and offer high-resolution graphics, preferably with good color capabilities. They should offer dynamic and animated displays and respond quickly when recomputation is required. Multitasking and window systems are highly desirable so students can work on mathematics and simultaneously have other tools, such as a word processor, available to support their work. They should provide networking for access to large file servers and expensive resources such as color printers, film recorders, or video systems. And, of course, they should be inexpensive.

Some of the more satisfactory computer systems that could be used for teaching mathematics include the Macintosh II, IBM PS/2 or PC/AT and compatibles, Sun 3, Apollo DN3500 or NeXT. However, these systems, adequately configured for visualization, are still too expensive for widespread use, and the initial NeXT system does not have color. The availability of appropriate software seems to determine if a system is used for education. This in turn is influenced by the ease of software development on the system, particularly during the first few years a system is available.

Software development is a problem for mathematics education throughout the entire range of computing systems. The graphics, windows, and sophisticated user interaction needed for effective educational software require a great deal of work to program. Professional programmers need a year or longer to become proficient on this kind of system, but faculty rarely have this kind of time to invest in building programming skills.

Issues in Visualization Technology

Both hardware and software technologies for visualization are developing rapidly. The funds and visibility the research community bring to this field support technological advances, and the large market for personal computers makes it feasible to put these advances into computers that education can afford. Systems now being used for research have interesting hardware developments which should be in personal computing systems in the next few years. These include faster processors and buffered displays to support more complex calculations and dynamic images, hardware 2-D and 3-D graphics accelerators, and very high-speed networks. Software for scientific visualization has advanced less than hardware, but full visualization systems such as Wavefront, NeoVisuals, and Doré are available, although these are often difficult to use. Various toolkits are being developed by the National Center for Supercomputing Applications and others. Most of these software developments are focused at the high end of research work, however, and have little impact on education.

There are some interesting developments underway in 3-D graphics. Some workstations can now present dual imaging with high-speed polarizing shutters and polarized glasses, allowing each eye to see a separate image and hence see a 3-D image. Less expensive video games offer polarized glasses synchronized with the display, which might be adapted to personal computers for displaying 3-D views. An approach without glasses uses barrier-strip technology pioneered by pscolograms from the Electronic Visualization Laboratory at the University of Illinois, Chicago; this technology is now in the laboratories. It is very important to distinguish between interactive, real-time 3-D and technologies such as computer-generated holograms, which allow the creation of 3-D examples but are not available for student explorations. In all these cases, there will be students who are not able to resolve a 3-D image because of problems with eye convergence; it is important that these students not be handicapped by an overdependence on 3-D technologies.

User interfaces for visualization are a continuing problem but work is going on in this area. Special attention is needed for interfaces to educational applications, since students will not have the time to become particularly skilled with any one system. General information on user interface options are beyond the scope of this article, but more information may be found in [2]. However, it is important to stress that user interfaces based on text-only computer terminals are now almost always unsatisfactory for educational work.

As a source of real-world input, few things can improve on video. Videodisks are an excellent way to provide images to students and instructors. These images can come from cameras, scanners, or remote sensing devices, such as satellites or space probes. Video can provide a point of contact with the real world as a background for a simulation or, if the images are matched carefully, as a check on the simulation's accuracy. Video images with suitable computer enhancement can also guide the student through an examination of complex structures or situations. Since video is such a new educational tool, we can expect to see exciting developments in educational video applications.

Incompatible systems pose serious difficulties for developers. Although work is underway to support common development for different kinds of computer systems, this may not yield significant help soon. For example, different windowing systems may produce very similar screens, but they require very different and separate development. The best suggestion for cross-system development is to use development systems available across a range of computers to build programs. System-independent software such as Objective C or the X Window System are the best kind of candidates for development across different machines. Obviously, software from hardware vendors is rarely portable.

Visualization in Mathematics

Visualization in the broadest sense has always been important in mathematics. In ancient Greece, geometers sketched diagrams in the sand; the calculus grew out of geometric efforts to solve problems of tangents, rates, lengths, and areas; and Hamilton invented quaternions as three-dimensional analogues of complex numbers which, taken as operators in three-space, rotate a given vector about a given line while scaling it. Moreover, even the vocabulary for communicating ideas is visual. Both in everyday life and in explaining mathematical ideas we say "Let me show you ...," and "Can you see ...?" We use blackboards or overheads in our lectures to draw diagrams illustrating our mathematics, and textbooks are often filled with diagrams. Computing has added to this visual environment as mathematicians have started to use computer graphics to explore or present their ideas.

Ancient mathematics depended only on number (arithmetic) and figure (geometry). Although it was possible to identify an algebra (which studied only numbers) separate from geometry by mid-Roman times, persons we would call mathematicians were called "geometers" as late as the 18th century [10]. As mathematics began to go beyond intuitive areas, however, visuals were seen as a weakness. For example, the deficiencies in Euclidean axiomatics were difficult to eradicate because the standard plane figures made the axioms "self-evident." The remarkable success of symbolic, formal mathematics in the late 19th and early 20th centuries left mathematics almost totally committed to symbolic work and tended to discredit visual approaches to mathematics.

Restoring the visual and intuitive side of mathematics opens new possibilities for mathematical work, especially now that computing has enough power and resolution to support it with accurate representations of problems and their solutions. The benefits of visualization include the ability to focus on specific components and details of very complex problems, to show the dynamics of systems and processes, and to increase intuition and understanding of mathematical problems and processes. The ability of some symbolic algebra systems to combine symbolics and visualization seems to offer computer-based mathematics the best of both worlds.

Visualization in Mathematics Curricula

Adding visualization to mathematics education promotes intuition and understanding and allows a wider range of coverage of mathematics subjects. But it provides more than this. Students not only learn mathematics but also learn new ways to think about and do their own mathematics. This is a key concept of educational visualization and meets one of the key problems in scientific visualization described in [7]: the lack of experience of scientists with visual understanding of their work. Visualization opportunities for science and mathematics education are described in [6] and [3].

Educational computing has only recently begun to work seriously on visual output. In the mid-1970's the author developed a visually-based computer laboratory for general statistics [4]; when it was submitted to a major developer of educational software, it was not possible to make the software portable and distribute it since the personal computer with graphics was not yet available. Many of the packages listed in the 1988 *College Mathematics Journal* database of educational software [5] did not use graphics. Most current visualization in educational software for mathematics grows from the textbook and blackboard diagrams we saw when we were students. These are limited uses of the graphics capabilities of computers, and we should be able to develop much more effective visual learning tools. However, this requires an understanding of visual learning that we do not have today. This is different from familiar symbolic learning in mathematics and needs a new set of communication tools and evaluation techniques.

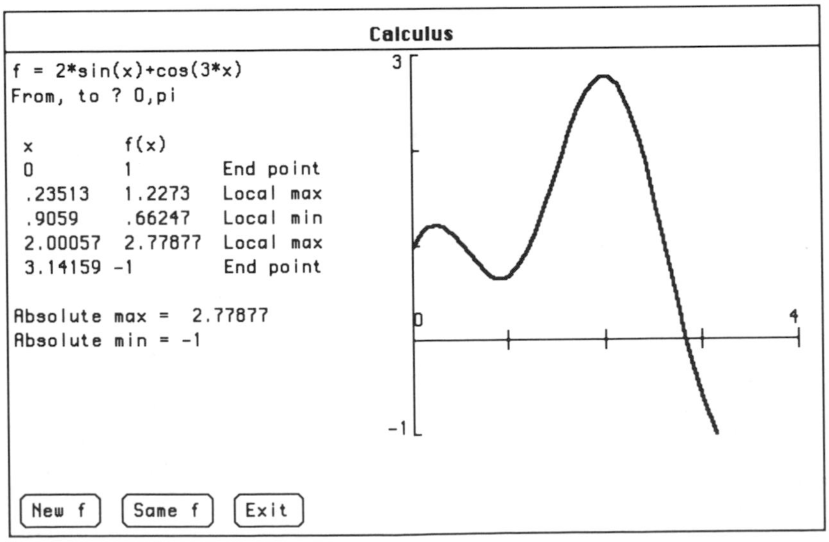

Mike Keeler, of Stardent Computers, describes three kinds of visualization. These are *postprocessing*, where the knowledge is complete and the user is creating a display of the finished product; *tracking*, where the knowledge is being developed and the user is watching it being displayed to see its nature; and *steering*, where the user is in the

Figure 1: *a plot of a function with maxima and minima computed, from the True BASIC Calculus package by John G. Kemeny*

```
                              Calculus
f = 2*sin(x)+cos(3*x)                      3
From, to ? 0,pi

   x        f(x)
   0        1          End point
   .23513   1.2273     Local max
   .9059    .66247     Local min
   2.00057  2.77877    Local max
   3.14159  -1         End point

Absolute max =  2.77877
Absolute min = -1
                                                          4

                                          -1

  [ New f ]  [ Same f ]  [ Exit ]
```

Figure 2: *a partly completed graph showing a cross-section through a developing surface, from FunPlot 3D by Walter Zimmermann*

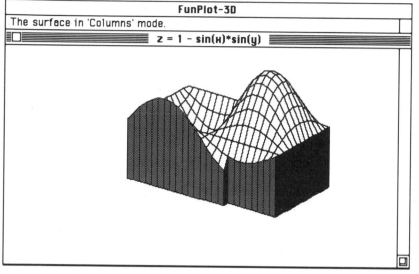

processing loop and can interact with and manipulate the simulation as it is under way. The Visualization in Scientific Computing report [11] emphasizes the need to steer calculations for scientific discovery.

Educational visualization can take advantage of these ideas. A simulation can present static images that report the results of a single computation, dynamic images that illustrate the behavior of a sequence of computations, or a steerable behavior that lets the student manipulate the simulation as it progresses.

Most educational visualization now uses postprocessing, since it focuses on presenting finished concepts to the student. This is illustrated in Figure 1, showing a function graph with computed maxima and minima. However, the author's informal studies with students across the sciences indicate that students respond much more strongly to dynamic images than to static ones. This suggests that presentations should use tracking or steering techniques whenever possible. These may involve showing a display as it is computed, which can be done so that the order of development illustrates the mathematical processes shown; for example, in displaying a surface it may be a good idea to plot back faces before you cover them up with front faces so the student will see that they are there, as illustrated in Figure 2, or a statistical experiment can show its samples as they are generated, as shown in David Griffeath's figure of the Galton board from his paper in this volume (Color Plate 9). These should be much more effective teaching techniques than simply using static images. Tracking can also be used in areas such as limit processes, where the intermediate computations leading up to the limit can be important to the overall process. Finally, steering techniques — actually getting the student involved in the development of a simulation, for example — are relatively unusual in educational software for mathematics, but may be found as trajectory traces for student-entered initial conditions in differential equations software as shown in Figure 3.

Figure 3: *a collection of trajectories for a differential equation with initial points in a square about the origin, from Differential Equations by Herman Gollwitzer*

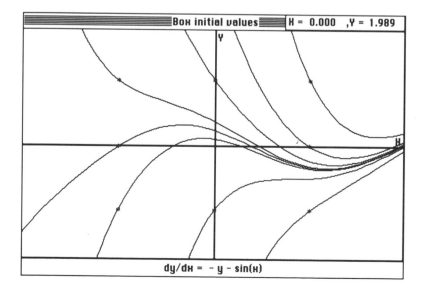

A different set of three visualization levels was described in the Visualization in Scientific Computing report [11]. These levels are phrased in terms of the audience for the images produced. The *presentation* level is intended to show work to a wide audience, for which all the tools of high-quality graphics are used; the *peer* level to communicate with others working at the same level, which require clear presentation but usually not the polish of the previous level; and the *personal* level to understand personal ideas in development, which may be quite rough and which need a good understanding of the problem to interpret the images produced. Figures 1 through 3 are probably peer level images, while Color Plate 1 shows a presentation-level image from George Francis and Donna Cox at the University of Illinois.

Educational visualization communicates to students who have some knowledge but who are not yet experts. Presentation visualization is too slow and takes too much effort to be effective for teaching, while personal visualization requires more knowledge of what is being displayed than most students can be expected to have. When presentation-level images are required, it seems most appropriate to use video or other precomputed images instead of interactive computing. An outstanding example of computer-generated video for mathematics teaching is the *MATHEMATICS!* project of Tom Apostol and Jim Blinn at Caltech; this project is creating a sequence of videotapes that teach fundamental concepts of geometry through innovative and dynamic presentations.

Another use of visualization techniques is to allow a student to experience different kinds of mathematical concepts and processes. These can be taken from traditional mathematical topics, such as topology, or can be new topics primarily accessible through visual methods, such as chaos theory. Kemeny [9] has described a number of ways computing is changing many courses. Taking advantage of visualization will require further work in curricular innovation and experimentation. Such innovation is now taking place within calculus, where the calculus reform initiative illustrates the kind of changes that will be needed to use computing and visualization effectively. As Barrett and Browder point out [1], the calculus changes "are not in the overall content of the course, but rather in the expectation of better student understanding; in greater stress on application and links with other disciplines; in the utilization of numerical methods and computer techniques; and in encouraging a fresh approach to teaching." Visualization also requires the developer and instructor to learn visual teaching and learning; these are not yet well understood.

It is well known that you cannot add something to a course or curriculum without giving up something else to make room for it. What do you give up to add visualization? It seems that manual symbol manipulation and calculation work are becoming less productive and appropriate than they once were; probably these are good candidates for replacement. Kemeny [9] has more detailed suggestions for particular courses.

Barriers to Visualization in Mathematics Education

There are several barriers to visualization in mathematics education. These lie in our institutional structures, in the cost of visual tools, in the lack of financial support for education, and in our understanding of teaching and learning. We do not see these barriers as insurmountable — in fact, they seem to be diminishing — but we believe it is important to understand them in order to make progress in the growth of this important educational technique.

Institutional barriers come from many sources, but the most critical is inertia in our institutions. It is difficult to get our departments (and ourselves) to adopt new teaching methods; it is difficult to adapt stable curricula to changes at all levels; it is difficult to get departments and others to accept curricular and instructional innovation as meriting academic rewards. In mathematics, there is also the question of getting the programs we serve to accept the validity of our course changes. Yet these are not insurmountable difficulties, because the paths of change are being made by the individuals who have written for this volume and by many others. This year there seems to be an upsurge of focused activity in visual approaches to mathematics, including a conference on visualization in research mathematics at the University of Iowa, an IMA conference on the mathematics of surfaces with research papers on " the visual recognition, manipulation, representation, design, and approximation of surfaces..." and a Chautauqua course on "Project *MATHEMATICS!*: A Case Study of Visualization in Mathematics Education." This indicates that interest in visual mathematics is growing in the entire mathematics community. There is even a forthcoming issue of *Leonardo* on images from visual mathematics, showing that the visual beauties of mathematics also are appealing to artists.

The cost of computing is still a problem for many mathematics programs. For years we took pride in requiring only paper, pencils, chalk, and a library to teach, and our administrations still expect mathematics education to make few financial demands. We now need to develop computing laboratories for mathematics programs. This means purchasing computers, developing networks, and providing technical support. We can use laboratories in engineering and computer science programs as models. However, the price of computing is still too high for schools with limited resources. We must look for grant or donation support, and try to get vendors to offer even more price reductions for systems whose use is limited to education.

Developing Educational Materials

Software development is another problem for education. Faculty members' primary dedication is to teaching in their disciplines, not to computing, and taking time to learn computing reduces professional opportunities in their own areas. One school of thought on software development holds that we need only develop a few versions of educational software for a mathematics course. If this is so, then traditional programming methods should be worthwhile. However, we believe that the diversity of educators' approaches to teaching should be reflected in the software used. This can only be true if educators can prepare customized computing experiences for their students. They must produce effective software quickly while ensuring that it emphasizes mathematical thinking. In a recent article [8], Steve Jobs of NeXT describes a typical faculty situation in developing an educational application by "I need it by next Tuesday and a grad student and I have to do it and we can (only) invest about two days in it." Many of us, alas, cannot even use graduate students.

One approach to developing course materials might lie in using sophisticated systems which allow the instructor to develop scripts to carry out specific operations. These are provided by *Mathematica* as well as other software. While this is a kind of programming, it is usually much less demanding than programming in traditional environments. If scripts can be changed by students, they can become actively involved in developing the mathematics involved, which can certainly heighten the educational effect. In any case, these systems remove the need for the instructor to program details of student interaction and displays.

Another approach could follow traditional software development more closely. Professional software developers spend a great deal of time in building toolkits for their applications. In the author's opinion, the best outlook for traditional academic development lies in adopting a similar approach using tools from Object-Oriented Programming (OOP) technology. OOP languages are specifically built to allow reuse of code and can include libraries which support many educational activities. For such systems, it is necessary to build the main objects that your software will use, and then to use these objects by adding the details of each application separately. Thus you do not need to rewrite the details of user interaction and graphic display, and you achieve software which works consistently with the student. The author would like to see professional development of nearly-functional educational programs with the user interface and display practically complete, so that an instructor or developer only needs to add specific details of user interaction and specific mathematical computations. This would allow an instructor to prepare an educational demonstration or simulation in a relatively short time without too large an investment in development.

Understanding Visual Teaching and Learning

Teaching based on visualization requires us to relearn many of our pedagogical skills. Not only must we understand mathematics, but we must understand how to communicate our mathematics visually. The authors in this volume are working diligently on this, and you will see many examples of such communication in these articles. There are many aspects of this visual communication. An instructor using visualization must:

- determine exactly the critical mathematical details to be presented in an image and show these either by highlighting them or by removing conflicting information,
- determine the order in which material is to be demonstrated by the images and present this material in a logical and connected sequence,
- offer students options in ways that expand their mathematical knowledge without confusing or overwhelming them,
- look for opportunities to present dynamic or developing mathematical processes and give students appropriate opportunities to explore or control them,
- consider carefully how students will learn visually, how to evaluate such learning, and how to integrate this learning with other parts of their mathematics studies.

One of the more difficult problems in adding visualization techniques to mathematics education is that we do not yet know how to evaluate this kind of learning. Visual learning does not provide familiar collections of mathematical facts or calculation processes. Visualization offers intuition, understanding, and concept formation which other disciplines evaluate by term papers and essay tests, hardly the kind of examination mathematics students are familiar with. Besides, standardized tests for graduate schools are not set up for this kind of evaluation, and neither are other standardized exams such as statewide competency tests. We need to learn how to make such evaluations a routine part of all mathematics examinations.

Probably the most important visualization issue for mathematics education is how visual and symbolic learning complement each other. At the actual course level, there are dual aspects to the problem of visual versus symbolic presentation. First, what material is best introduced visually, and what is best introduced symbolically? For material that needs a visual presentation, when (if ever) should this be treated symbolically? What makes the visual treatment eventually inadequate and requires symbols? Perhaps it is a need to get more precision or accuracy, to acquire a vocabulary to do work that goes beyond standard images and two- or three-dimensional displays, or to develop treatments that apply broadly to a wide range of topics. Conversely, for material that is best introduced symbolically, when (if ever) should this be treated visually? Perhaps visualization gives added value to these topics through development of intuition, an overall understanding of a complex problem, or the ability to see the topic's principles in an application. For example, *Mathematica* notebooks merge textual, symbolic, and graphical information in a way that allows the individual student to explore each in the order that fits his or her curiosity and learning needs best.

These questions merit focused research in their own right. At this point, there are many kinds of development which provide potential models for visual learning, but these come from individual developers' tastes and teaching styles. We need more examples of visual education in mathematics, some classification of the visual approaches used, and formal evaluations of these various visual techniques. Finally, research must be done on structured testing and evaluation of visual learning and we must build working models of such evaluation, so the agencies which use these evaluation tools will be able to consider visual learning properly.

Conclusions

Computing technology is making it much more rewarding for mathematics to use graphics, and in turn mathematics is showing an increased interest in visual approaches to both teaching and research. This offers the prospect of increasing student understanding of mathematics and mathematical processes. However, taking full advantage of this technology will require work from the mathematics community in understanding visual teaching and learning, evolving course design, new methods of evaluation, and further technological development.

References

[1] Barrett, Lida K. and William Browder, "Reflections on the Calculus Initiative," *UME Trends*, 1(October 1989), 8.

[2] Brown, Judith R. and Steve Cunningham, *User Interface Programming: Principles and Examples*, Wiley, New York, 1989.

[3] Brown, Judith R. and Steve Cunningham, "Visualization in Higher Education," *Academic Computing*, March 1990.

[4] Cunningham, Steve, "A Computer-Based Laboratory for General Statistics," Proceedings of the National Educational Computer Conference, 1979, 291-296.

[5] Cunningham, Steve and David A. Smith, "The Compleat Mathematics Software Database," *The College Mathematics Journal*, 19(May 1988), 268-289.

[6] Cunningham, Steve, Judith R. Brown, and Mike McGrath, "Visualization in Science and Engineering Education," in *IEEE Tutorial: Visualization in Scientific Computing*, Gregory M. Nielson and Bruce Shriver (eds), IEEE Press, 1990.

[7] DeFanti, Thomas A., Maxine D. Brown, and Bruce H. McCormick, "Visualization: Expanding Scientific and Engineering Research Opportunities," *Computer*, 22(August 1989), 12-25.

[8] Denning, Peter J. and Karen A. Frenkel, "A Conversation with Steve Jobs," *Communications of the ACM*, 32(April 1989), 436-443.

[9] Kemeny, John G., "How Computers Have Changed The Way I Teach," *Academic Computing*, 2(May/June 1988), 44.

[10] Kline, Morris, *Mathematical Thought from Ancient to Modern Times*, Oxford University Press, New York, 1972.

[11] McCormick, Bruce H., Thomas A. DeFanti, and Maxine D. Brown (eds), Visualization in Scientific Computing, *Computer Graphics*, 21(November 1987).

Acknowledgement

This article is based partly on work supported by a grant from the California State University Research, Scholastic, and Creative Activities program. It was developed in parallel with [3] and shares a number of ideas with that paper.

The Difference Between Graphing Software and Educational *Graphing Software*

E. Paul Goldenberg

From the time that students have even the crudest idea of what a functional relation is, the thoughtful use of graphical representation can be of educational value to them. There are now quite a few software tools and even calculators that reduce the overhead involved in producing graphs, making it possible for students to explore a greater number of graphs than they could if they had to graph by hand. In the face of such opportunity, it becomes important to ask what special considerations must go into the design of a graphing tool for educational purposes. In what ways should such a tool differ from one that is suited for engineering or scientific purposes?

There are reasons why some differences might be expected. For one thing, when engineers and scientists use graphers, they are often interested primarily in the behavior of a *particular* function. Although students, too, must deal with particular functions, most of the educational value is in the generalizations they abstract from the particulars. The shape of $-2x^2+30x-108$ is of *no* educational consequence, but it may serve as a data point about any of several broad classes: a particular family of quadratics (e.g., ones that differ only in the constant term); more generally, all quadratics; still more generally, all polynomials or even all functions. This difference in purpose has implications for the user interface.

Lesson 1: It must be easy to modify functions.
For educational purposes, the interface must make it convenient to perform experiments in which students modify a single parameter (coefficient or exponent) within a given form in order to study the effect.

Lesson 2: It must be easy to compare functions.
For students to make effective comparisons of different forms, such as comparing $2(x-3)(x+5)$ with $2(x+1)^2-32$ and $2x^2+4x-30$, or comparing the graphs of two functions with the graph of their sum, difference, product, quotient, or composition, they must be spared the time it takes to retype each form. They must also have opportunities to superimpose graphs and to view graphs in multiple windows.

There is a second reason why we might expect there to be different needs in educational and scientific graphing. We all bring a variety of general strategies to our interpretation of graphs, but research shows that we must also bring specific mathematical knowledge and expectations if we are to interpret graphs correctly. Interpretation becomes particularly important when the emphasis is not on reading specific values off of the graph, but on abstracting and relating features of several graphs. Students who lack such

E. Paul Goldenberg received his doctorate in curriculum and supervision from Harvard Graduate School of Education. He is currently a Senior Scientist at Education Development Center, Inc. (EDC) in Newton, Massachusetts, holds an appointment as Associate in Education at Harvard Graduate School of Education, and is an adjunct faculty member at Lesley College in Cambridge, Massachusetts. He is author of Special Technology for Special Children, principal author of Computers, Education, and Special Needs (Addison-Wesley), and Exploring Language with Logo (The MIT Press), and editor of The MIT Press series Exploring With Logo. He directs research in visualization in mathematics and collaborates in the development of curricula in mathematics and linguistics.

mathematical knowledge often misinterpret what they see and may invent complex and misleading explanations for the interrelatedness of graphs (Goldenberg, [2]). There is a great deal of subtlety to the interpretation of graphs, and much is still unknown about how either experts or students do it. Still, each understanding that we gain shows its practical value by allowing us to build tools and surrounding curriculum in a way that either eliminates the confusions or uses them to educational advantage. Recognizing patterns in students' misinterpretations can help us generate mathematically rich problem situations that derive from the ambiguities in graphic representations of functions and aim ultimately to overcome them (Goldenberg, [2]; Harvey, et al., [3]; Mark, et al., [4]).

■ **Figure 1**

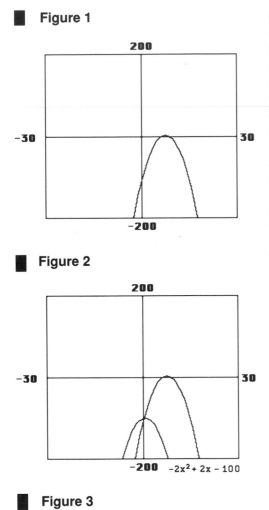

Goldenberg [2] reports the work of two bright, successful, second-year algebra students who experimented with various quadratic polynomials to match a target graph whose symbolic representation ($-2x^2 + 30x - 108$) was unknown to them (Figure 1). They were encouraged to use whatever means they chose, including making various computer-supported measurements on the graph, trying out expressions, and observing differences between the graphs they created and the target at which they were aiming.

On the basis of prior exploration with quadratics in polynomial form, the students had built up some expectations about the effect of the three coefficients on the graph. Their experiments showed that the quadratic coefficient controlled something they referred to as "shape," though they were not completely certain whether the parabola grew "fatter" or "thinner" as the magnitude of the coefficient was increased. They knew and were certain that, because the target parabola was "upside down", its quadratic coefficient must be negative. After a single experiment they reasoned from the "pointiness" of the parabola that the x^2 coefficient might be around -2. They also had a notion of "height" and believed that it was controlled by the constant term. The measure of "height" that they chose was the y-intercept, which they estimated by eye to be about -100. They stated that they had no idea what to use for the coefficient of x, and they arbitrarily picked 2.

■ **Figure 2**

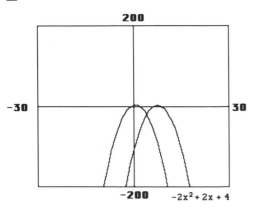

Figure 2 shows the function that they designed, $-2x^2 + 2x - 100$. From *our* perspective, their analysis (at least as far as it had gone) had been excellent. As well as one can tell at this scale, they seem to have matched shape and y-intercept perfectly, leaving them only to discover the x coefficient in their next experiments. But *their* perspective was very different. Because their parabola *looks* lower than the target, their attention was drawn away from the measure of "height" that they had originally chosen (y-intercept). In their view, they had gone wildly wrong, failing even to find the constant term, and so they first "corrected" the constant term from -100 to +4, to bring the level of their parabola up to the same "height" as the other. The new expression, $-2x^2 + 2x + 4$, was in some ways further from the target than the first, yet it was more satisfying because it was just as "high" as the graph they were aiming at (Figure 3).

■ **Figure 3**

In this case, the confusion appeared to result from a shift of attention from one feature (the y-intercept) to another (the overall "height" as suggested, perhaps, by the height of the vertex). In other cases, the confusion seems to arise from mechanisms that give rise to well-

■ Figure 4

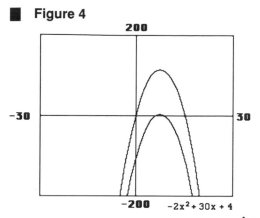

$-2x^2 + 30x + 4$

known perceptual illusions. For example, when these same students begin to work on the linear coefficient, expecting that it would control horizontal position, they produced Figure 4 in which all but the constant term (the term they just altered) is correct. The students knew that they needed to adjust the constant term, but they also remarked that the inner parabola looked more obtuse than the outer. (In this example, the target parabola appears blunter than the upper one.) As was true of their confusion regarding the meaning of "height," illusions such as the one illustrated here were sometimes powerful enough to draw their attention back to the already correct coefficient of x^2 and cause them to change it. We may draw at least four lessons from this single example:

Lesson 3: Certain regions and markings draw attention.
Students base comparisons of graphs on gestalts that they create by matching identifiable "special" points or regions such as, in this case, the vertex of the parabola. To help counter the attention that the vertex draws, it may be important to mark, or better yet, let students mark, points whose transformations they wish to track. One such point might be, for example, the *y*-intercept.

Lesson 4: Explanatory language is important.
The two parabolas in Figure 2 *are* graphed at different "heights" and the constant term *does* control this height. But to *compare* two parabolas and know something about the relation of their constant terms, we must look only at "height at $x = 0$." Alternatively, we must avoid using the term *height* and stick strictly with *y-intercept*. Language being what it is, it would be unnatural (and probably ineffective) to prescribe usage, but the issue should be made clear by drawing explicit attention to it.

Lesson 5: Students need to modify graphical representations.
The perceptual illusion illustrated in Figure 4 results from the visual strategy of measuring distance from a point on one curve to the *nearest* point on the other. Since these distances are not constant, the parabolas do not appear to be congruent. When students drag a parabola vertically with a mouse (or move it vertically stepwise with an arrow key), the illusion vanishes, as they tend to measure distances along the direction of movement rather than to the nearest point.

Lesson 6: The wording of the problem influences the solution
Our students faced the task of determining the three parameters *a*, *b*, and *c* in the expression $ax^2 + bx + c$. They discovered that *a* fixed the shape of the parabola. All that was left for the others to do was fix the position. Because they knew that *c* moved the parabola vertically, it was quite reasonable to assume that *b* moved it horizontally. This assumption got our students into trouble, but it would have been perfectly appropriate if the form in which they were to express their results had been $a(x - b)^2 + c$. There is no best form, nor can we exhaust the set of essential forms. Curriculum must therefore draw attention to the issue: the *form* of a problem influences how we think about it. Altering the form of a problem can help us solve it. We should design curricula to make it very apparent to students that the purpose of an algebraic manipulation on some expression is quite often to create a tautology: a new expression representing the same underlying object but changed in form so as to reveal some property of interest.

Lessons 3 through 6 were drawn from an example that only begins to suggest the difficulty students seem to have in keeping track of individual points when they face the overall gestalt of the continuous curve. It is quite evident from our other work that this is a major problem for students. For one example, students' ability to use graphs successfully to solve simultaneous linear equations was influenced by how they created the graphs. From the comments of some students, it appeared that when they plotted the graphs from tables, they easily recognized that more points might fall between the entries on their tables, and so if the common point did not appear at the (nearly always) integer x values they tried, a solution in between those points would be tried. When they used the slope-intercept method, many evidenced a different sense: their drawing consisted of two kinds of geometrical objects, one line and two points. They seemed to lose the sense that there were any other points on the line (even though they could *put* a point anywhere). We theorize that this sense is related to another problem, in which students fail to distinguish the status of x from that of a, b, and c in the definition $f(x) = ax^2 + bx + c$. All of the letters are stand-ins for numbers that may take on any value, but students fail to see that the left-hand side of the definition indicates that once *any value* is selected for each of a, b, and c, the function describes a mapping for *all values* of x in the domain. Instead, students seem to see the products of graphers as gestalts, not as $\forall x$ mappings, just as the slope-intercept students saw lines instead of sets of points, as if they were dealing with a function of three variables $f_1(a,b,c) = ax^2 + bx + c$ that produced a graphical object as its output.

This sense is apparently aided by the inexorable (and therefore invariable) left-to-right sweep of conventional graphers and the quite variable nature of the parameters one may enter into the graphable function. In other words, from the students' perspective, they do not vary x, but do vary a, b, and c. Many possible technological as well as curricular strategies may aid in resolving this problem.

Lesson 7: Help students see discrete points in the gestalts.
One possible strategy is to eliminate the inexorable left-to-right continuous sweep. If all points in the visible domain were plotted in random order, students would see discrete examples of the mapping before they saw them fuse into a form. The effect would suggest an "if-x-is-this-then-$f(x)$-is-that" image, rather than a geometric shape which swallows up the points and, even more, totally swallows up their symbolization of pairs of numbers. Another strategy would include tables of values. A third strategy would highlight the point-set nature by lighting only every other pixel (that is, never developing the continuous *image* but forcing students to imagine it). A fourth strategy would require students to use a mouse to click at individual x "points" or sweep over x regions that they wish to have plotted, putting the variable back under their control. Each of these strategies may have its own disadvantages. (The third strategy, in particular, leaving a dotted line for the function, may play into students' all too strong sense that points are like physical objects.) Research needs to be done to see which (combination?) of these (or other) images along with what pedagogical strategy makes for the clearest understanding of variable, the continuous nature of number, and the $\forall x$ notion.

Our concepts of what is elementary and what is hard may need a radical readjustment. Typically, we teach about linear functions before we teach about higher order polynomials, transcendentals, and other "advanced" functions. But that order is conditioned at least partly by the simplicity of the *algebra* and the ease of performing numerical computations. When more emphasis is placed on graphical representations — in particular, on *comparing* graphs and abstracting their relationships — the linear functions are *much* harder to interpret than functions that have more visually distinguishable shapes.

Consider, for example, an assignment in which students are to investigate the effects of the constant in the linear form *ax+b* by experimenting with many graphs and abstracting a rule. Left to their own devices, students tend to use positive integer values in their explorations and so are likely to encounter graphs like the pair of linear functions *f* and *g* illustrated in Figure 5. As a result, unless they have some prior mathematical expectations (which would also make the experiment unnecessary), they are not likely to perceive *g* as higher than *f* but rather see it as sitting to the left of *f*. And, of course, such a mapping *is* perfectly reasonable though it is educationally confounding. In the form *m(x-r)*, the *r* would most sensibly be interpreted as controlling the horizontal position, but multiplying through gives the expression *mx-mr* where *-mr* is the equivalent of the *b* in the earlier form. For a fixed slope and varying *y*-intercept, we now seem to be making contradictory claims: *b* controls vertical position, but *b* times a constant (−1/*a*) controls horizontal position.

Figure 5

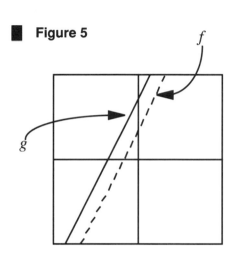

Though the apparent direction of movement is a perceptual phenomenon that no amount of algebraic sophistication can change, algebraic sophistication can lead us to ignore appearances. Students need other aids. They may be instructed to use the *y*-axis as a guide, but that seems a formulaic and rigid way to help convince them that the movement is vertical even though they can see, not subtly but glaringly, that it is horizontal. Lesson 5 offers a better alternative: the experience of dragging the graph vertically with a mouse and watching the algebraic expression change can help attract attention to the direction of movement and help students see *in their heads* that the *segment* of the infinite line representing *f* is mapped as in Figure 6. (However, building the imagery of Figure 6 into the graphing tool would be inappropriate unless the domain of the function were restricted as the figure suggests.)

Figure 6

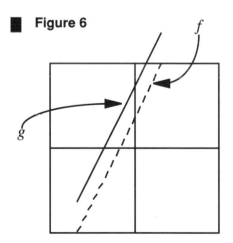

Lesson 3 also offers a strategy: tic marks and labels on the graph can help students interpret the graph in a useful and unambiguous way. The addition of one or more points, e.g., *f* (0) and *f* (1), helps the student trace the movement of the entire graph by providing visual cues that lead them toward a vertical rather than horizontal interpretation. But beyond these strategies for getting students past the difficulties, another lesson emerges.

Lesson 8: Linear functions may not be the way to begin.
What is easy and what is hard depends on the tools one uses. When approaching functions through their graphs, it may make most sense to begin with graphs that have *no* convenient algebraic representation and with notions that we typically ignore until the calculus, including the nature of the domain, local maximum or minimum, rate of change, and continuous or abrupt change. An interesting implementation of this strategy, though one that makes minimal use of the computer's power, is *Interpreting Graphs* [1]. When dealing with algebraic functions, it may still make sense to hold linear functions for last.

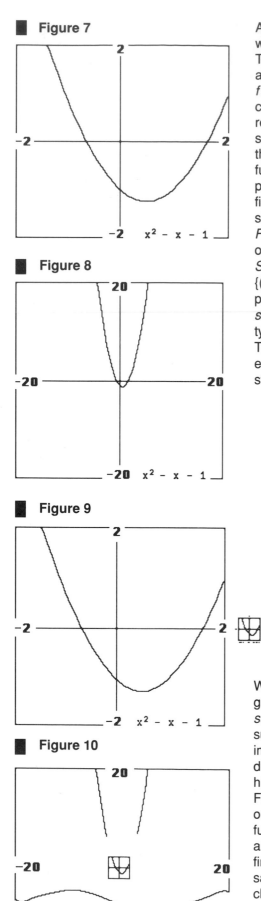

Figure 7

$x^2 - x - 1$

Figure 8

$x^2 - x - 1$

Figure 9

$x^2 - x - 1$

Figure 10

An increased use of graphing may require us to rethink the order in which concepts unfold in the algebra curriculum in other ways as well. The subtleties in Figures 5 and 6 alone suggest a new approach. First of all, let the dotted line in Figures 5 and 6 represent $f: \Re^1 \to \Re^1$ defined by $f(x) = 2x + 2$. Traditionally, and with little acknowledgement of the confusions or misconceptions that may arise, students are asked to represent this $\Re^1$ to $\Re^1$ mapping with points in $\Re^2$. Then, experiments such as those suggested in the figures involve $\Re^2 \to \Re^2$ mappings, which the students most likely have never encountered in a formal way. To further confuse matters, the two transformations — the ways that the points in $\Re^2$ are mapped from the dotted to the solid lines in the two figures — are quite distinct, but their *results* are identical. Figure 5 shows a mapping that could be represented by $F: \Re^2 \to \Re^2$ defined by $F(x,y) = (x-1,y)$; Figure 6 shows $G: \Re^2 \to \Re^2$ where $G(x,y) = (x,y+2)$. It is only when F and G are applied to certain point sets $S_c = \{(x,y) \in \Re^2 \mid y = 2x + c\}$ that the resulting point sets are identical: $\{(x,y) \mid y = 2x + c + 2\}$. Applied to other point sets, F and G will produce nonidentical results. Thus the *lines* represented by the solid *segments* in Figures 5 and 6 coincide because f generates a point set of type S_c, specifically, S_2. It is not just my notation that makes this hard. The difficulty is that so many different ideas, all new to the student, are encountered at the same time. How would we ever expect a beginning student *not* to be confused by all of this?!

Lesson 9: $\Re^2$ representations of $\Re^1$ to $\Re^1$ mappings may not be the way to begin.

This is an even more radical departure from tradition. Although it is unreasonable to expect beginning students to understand the subtleties of $\Re^2 \to \Re^2$ transformations performed on $\Re^2$ representations of $\Re^1 \to \Re^1$ mappings all at once, the separate steps are quite reasonable to teach. We might begin with $\Re^1$ representations of the $\Re^1 \to \Re^1$ mappings: parallel "input" and "output" number lines with an animation that shows the movement of the image on the range line as the input is moved, mouse-in-hand, on the domain line. A subsequent perpendicular arrangment of these number lines gives a new image, especially if the vertical range (output) line is allowed to move horizontally, always passing through the input point on the domain line. A separate game of $\Re^2 \to \Re^2$ transformations performed on points *always well understood to be members of* $\Re^2$ makes sense before exploring $\Re^2 \to \Re^2$ maps on $\Re^2$ representations of $\Re^1 \to \Re^1$ maps.

When we interpreted Figure 5, we treated f and g as different functions graphed at the same scale. Though it generally makes little sense to *superpose* two views of the same function graphed at different scales, a superposition of f with a "close-up" (labelled g) would produce the same image. Such a linearly symmetric scale change — a "zoom" — can even distort apparent shape, rendering incomplete the observation that the high-order coefficient fixes the visual shape of the graph of a quadratic. Figures 7 and 8 are close-up and distant views of the same function, yet our eyes tell us that they are different "shapes" and so must be different functions. The numeric labels on these graphs indicate their scale, but are insufficient to convince us that the functions are the same. Yet we find it quite easy to believe that the two images in Figure 9 show the same object at two different levels of magnification. When the window changes along with the scale of the object in it — that is, when we see the frame as well as the parabola from afar — the two views appear to show the same curve. Figure 10 shows how the smaller image in Figure 9 relates to the graph shown in Figure 8.

Lesson 10: Scale information needs to be presented visually.
When multiple scales are used to represent the same graph, graphing windows should contain internal frames or other visual aids to help students recognize which portion of a distance view is being enlarged in a close-up view. (The "launch" and "put away" animations on the Macintosh, with rectangles expanding or contracting rapidly and more-or-less in the right direction, seem suggestive.) Numeric labels are not enough.

Although asymmetric scales might seem an unnecessary complication to add to a student's life, they are in fact inescapable if students are to use graphing software flexibly to perform their own experiments. Even with linear functions, extreme slopes will require an asymmetric scale to give a useful view of the graph, and the need for asymmetric scale becomes greater and greater as the degree of a polynomial increases. Sometimes an understanding of the effects of scale is necessary in order to interpret apparent anomalies in the graph. At high compression, a local feature (e.g., a local extremum) may fall between adjacent pixels and be missed altogether, generating a vastly misleading image on the screen.

The problem of pixel approximation might well be a valuable vehicle for instruction in a curriculum designed to take advantage of it. On the other hand, in a curriculum not so adapted, it may create more confusion than clarity. As with other issues, this one suggests that curriculum must not be ignored.

Lesson 11: Scale needs to be taught.
The scale issue cannot be avoided in a computer graphing environment. Nor should it be, as there is important mathematics in realizing that the graphs of $(x + 10)^3 - 40x$ (big "bumps" displaced to the left of the origin) and $(x - 10)^3 + 40x$ (no "bumps" at all, and displacement to the right of the origin) can be made to coincide visually with each other *on the same computer screen at the same scale*, both looking like the canonical representation of x^3 (with which they will also coincide). Again, the mathematics is generally not touched until the calculus, but states that the global behavior of these functions — the limiting behavior as the magnitude of x grows — is dependent on the highest order term and not on the "details" buried in the other terms.

Lesson 12: Provide quantitative *and* qualitative scale controls with no bias toward symmetry and linearity.
It is important that students have experience *controlling* scale. In fact, students need experience both with strictly metric controls (e.g., specifying the exact borders of the region of the plane they wish to examine) and with visual, primarily nonmetric controls (e.g., stretching or shrinking an image to reveal or match some property without having to compute window border values in x and y).

The default behavior of the controls should foster a richer rather than poorer view of the space within which the functions reside. In particular, "zoom" functions suggest a linear relation between the expansion in the two directions. When we zoom in on a graph to examine local behaviors such as roots, tangents, and local extrema, the results make sense only if the graph is *locally linear*. While this corresponds well with our own visual experience, it works against "seeing" the solution to the problem of getting the cubics in Lesson 11 to coincide: the results of zooming *out* are visually confounding unless the function is *globally linear*. Such expansions might reasonably be tied, in the case of polynomials, to the degree of the function. Scale controls should not reinforce students' already existing preferences for symmetry and linearity.

Because a visual approach changes what is and is not accessible to students, thought must be given to the kinds of educational questions that we ask in the new environment. Not only does rapid graphing make new questions possible, but it makes some new questions, e.g., ones that draw attention to the possible ways of misinterpreting the newfound graphic information, quite important.

Knowing that graphing packages can invite misinterpretation, and knowing even what directions the misinterpretations are likely to take, we are obliged to develop less ambiguous presentations if we intend to use them for educational purposes. We must also treat the potential misinterpretations of graphs as challenges to explore the mathematical richness behind them. Not only must the questions be raised, but class time must be allocated for discussion of students' observations and conjectures.

It is quite plausible that a more visual approach to algebra will give rise to a style of mathematically solid visual reasoning that is quite accessible to students who found the symbolic/syntactic techniques cumbersome or devoid of meaning. I will illustrate this idea with a problem and a visual argument for its solution.

> **Problem:** Consider the family of parabolas ax^2+bx+c where a and c are fixed and b varies. Along what mathematical curve do the vertexes of these parabolas lie?

Conventional approaches begin by finding the vertex either through the calculus or using more advanced algebraic techniques such as completing the square. A little experimentation, however, can lead one to a visual attack on the problem that requires almost no algebra at all.

Students observe that the coefficient a of the quadratic term controls the "shape" of the parabola (on a fixed scale). The greater a's magnitude, the pointier the parabola. (Having made this observation, students can then prove it is true through essentially visual and qualitative argument, but I will forego that proof here for brevity.) Since a is a fixed quantity, the "shape" of the parabola is fixed. Let us embody this idea concretely, building the viewable portion of the parabola out of stiff coat-hanger wire (Figure 11).

Students also observe and prove that the constant c determines the y-intercept. Because c is fixed, we may embody the idea of the fixed y-intercept concretely by picturing it as a screw-eye through which the wire-parabola must pass (Figure 12).

Now, all that is left for b to control is what portion of the parabola passes through the screw-eye. Experimentation with the graphing software *shows*

Figure 11 **Figure 12** **Figure 13**

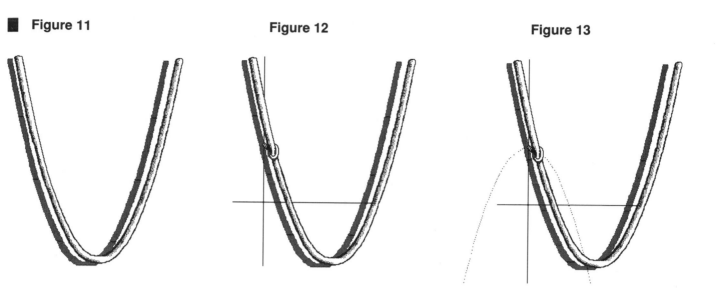

that the locus of the vertex points traces out a curve that appears to be an "upside-down" and "centered" copy of the moving wire-parabola. Guided by that expectation, one may reason from the concrete image, arguing that as the wire-parabola (always retaining its vertical orientation) is slipped back and forth through the screw-eye, the vertex must trace out the parabola's shape and will be at its maximum height at the screw-eye (Figure 13). So we conclude, informally, that the curve $-ax^2 + c$ passes through the vertices of all the parabolas in this family. If a formal symbolic proof is desired, this reasoning has led us to the proper conjecture for proof and provides a valuable scaffolding for the required symbolic manipulations.

Lesson 13: Curriculum must accommodate visual thinking.
At the least, we must explore the new problems that students can pose and solve in this new environment. More radically, we must explore what changes in the overall curriculum make sense. We have already suggested the presentation of nonlinear functions before linear ones and an earlier introduction to transformations on the plane. A visual/intuitive calculus before most of algebra also seems highly plausible and worthwhile. And an introduction to some topological ideas may make sense as part of the study of functions.

References

[1] Dugdale, S. and D. Kibbey. *Interpreting Graphs*, Sunburst, Pleasantville, NY.
[2] Goldenberg, E. P., " Mathematics, metaphors, and human factors: mathematical, technical, and pedagogical challenges in the graphical representation of functions," *Journal of Mathematical Behavior*, 7(1988), 135-174.
[3] Harvey, W., J. Schwartz, and M. Yerushalmy, *Visualizing Algebra*, Sunburst, Pleasantville, NY, 1988.
[4] Mark, J., M. Yerushalmy, and W. Harvey, *Visualizing Algebra: The Function Analyzer Problems and Projects*, Sunburst, Pleasantville, NY, 1989.

The Software

Prototypes of two pieces of software developed at Education Development Center, Inc. were used in this research: *The Function Analyzer* and *The Function Supposer: Explorations in Algebra*. Final versions of these pieces were informed by this research. *The Function Analyzer* enables one to explore and manipulate linked graphical, symbolic, and tabular representations of functions. It allows, among other manipulations, direct *graphical* modification of functions through translation, dilation, or reflection. *The Function Supposer* extends this environment and enables one to construct functions through arithmetic operations (and composition) on other functions. Both pieces of software are from the *Visualizing Algebra* series and are published and distributed through Sunburst Communications.

Acknowledgements

This report and some of the ideas behind it were supported in part by the National Science Foundation, grant number MDR-8954647. The views represented here are not necessarily shared by the NSF. Special thanks are due to Philip G. Lewis for his insights throughout, and particularly for his ingenious imagery reflected in Lessons 7 and 9. I am also indebted to Wayne Harvey and Judah Schwartz for substantial intellectual contributions to this work.

This paper was originally prepared for the second annual Conference on Technology in Collegiate Mathematics, November 3, 1989, and will appear in the *Proceedings* of that conference (F. Demana and B. Waits, eds.), Addison-Wesley, 1990. It is copublished here by request of this volume's editors and with permission of F. Demana, B. Waits, and the author.

STEREOMETRIX - A Learning Tool for Spatial Geometry

Tommy Dreyfus and Nurit Hadas

Do the four altitudes of a tetrahedron meet in a single point? One of the difficulties in answering this and other questions in stereometry, or three-dimensional Euclidean geometry, is our lack of ability to visualize complex three-dimensional objects.

Spatial geometry used to be a compulsory topic in high school. Today it is at best optional. In the Israeli high school curriculum, for instance, only students learning mathematics at the highest possible level deal with this topic seriously; in France, it is considered "the 'poor relation' of the teaching of mathematics" [7]. Spatial geometry is, however, important, at least for future scientists, engineers and technicians. The former need it in model building, the latter in the design of machines.

Learning Difficulties

Solving stereometry problems requires thinking in two representations, visual and analytic; the combination and integration of these are not easy and need serious attention in the teaching process. Although specific learning difficulties in stereometry have not been investigated, there is some research work on the understanding of the transition between three-dimensional objects and their two-dimensional representations; this transition is bidirectional: imagine, and draw projections of a given solid and interpret a plane figure as the projection of a three-dimensional solid. Much of the research on this transition dealt with polycubal solids and students' ability to create and interpret two-dimensional representations for them; in most cases, it was found that students had great difficulty in successfully communicating visual information as well as in interpreting two-dimensional drawings of polycubal solids (for a recent review see Section III of Hershkowitz [3]). Parzysz [7, p. 79], on the other hand, investigated students' implicit rules used in the transition between two and three dimensions in the case of a square base pyramid. He reported that his students tended to confound the three-dimensional figure drawn with the two-dimensional one having the same representation. In drawing, the main difficulties he found originated from a conflict between what is seen and what is known; for instance, the base of the pyramid was more often rendered as a square or diamond than as a rectangle or parallelogram.

Students generally have a tendency to base arguments on appearance. For instance, an angle which in the three-dimensional solid is a right angle may appear acute (or obtuse), and the opposite: an acute angle may appear in the projection to be a right angle. In fact, the following statement, which may appear surprising, is true: let α be any angle $0 < \alpha < 180°$; let β be another

Tommy Dreyfus is Associate Professor of Science and Mathematics Teaching at the Center for Technological Education (affiliated with Tel Aviv University) in Holon, Israel, and is also a consultant at the Department of Science Teaching of the Weizmann Institute of Science in Rehovot, Israel. His main research interest concerns the effects of computer use on students' mathematical thinking patterns.

Nurit Hadas is a senior Curriculum Developer at the Department of Science Teaching of the Weizmann Institute of Science in Rehovot, Israel. She has participated in projects developing courses in geometry, statistics, and algebra, mainly functions. She has also developed several software programs.

such angle. Then there is a projection of the plane containing α such that the projected α equals β. Similarly, a regular tetrahedron appears, from certain angles, just like a general one and the opposite: many general tetrahedra may be made to look rather regular. Correct interpretation of two-dimensional drawings of solids is one of the principal aims of teaching spatial geometry.

Computer Tools

To what extent can computers be used to help with the processes of visual representation and interpretation necessary in geometrical problem solving? Plane geometry has been a trial ground for this (see, e.g., [5], [2], and [8]). Hoyles [4] has noted that the computer's role in this situation is to be a " tool for the manipulation of graphical representations which has the power to provide informative feedback" (p. 66). In stereometry problems, the students' reasoning can be based on their interpretation of the visual representation in addition to the analytic information. Visual support is particularly important in this area because of the difficulty with two-dimensional representations mentioned above.

Osta [6] used the dynamic potential of the computer to teach spatial topics, specifically polycubal solids, to grade 8 and 9 students and to observe their progress. She found that in the course of the instructional sequence, they passed from using more perceptual criteria to using more geometrical ones.

Although the representation of a solid on the computer screen necessarily remains two-dimensional, this representation is dynamic. This added flexibility can be used to
- have the constructions executed by the computer. This often forces the student to base his work on definitions and thus prevents "cheating" by imprecision and wrong pictures. This occurs for instance with angles, say between planes, which can not be simply drawn but must be constructed.
- make students' misconceptions explicit and thus create a conflict,
- prevent mistakes during the transition between two- and three-dimensional representations, and
- lead the students to conjecture properties and check these conjectures with the help of the computer. This may again provoke conflict between what is expected to happen on the basis of unwarranted assumptions and what is actually happening on the screen.

These considerations guided the design and construction of STEREOMETRIX, a learning environment for stereometry, in particular for geometric constructions on standard solids such as cubes, pyramids, prisms, cylinders, and cones.

The Software

The STEREOMETRIX software is operational in the sense that the students act by operating on mathematical objects. There are two main types of operations: constructions and transformations. Constructions are operations that change a given solid. Transformations are operations that do not change the solid but change its position in space and thus its representation on the screen.

Constructions. The constructions listed in Figure 1 can be carried out on a solid after it has been chosen from the list of available solids. (One of the

Figure 1: The available constructions.

```
CONSTRUCT
>Perpendicular
 Bisect angle
 Compass
 Intersection
 Connect
 Disconnect
 New point
 Move point
 Erase point
 Rename point
 Mark angle
 Quit
```

available "solids" is the empty space, an option which allows the user to construct his own solid from scratch.) Some of the constructions which can be carried out automatically are perpendiculars from points to either a line or a plane, and copying segments. Thanks to the computer, the resulting representation of the solid depends neither on the students' ability to draw (see Example 1) nor on their misconceptions (see Example 2).

Transformations. The software enables the student to rotate, translate, and scale solids. Rotation is possible around any of the three axes *x, y, z* (a small coordinate system always appears on the screen) in order to turn the solid into a convenient position. A paper and pencil drawing is always single, isolated and static. The student's ability to rotate a three-dimensional solid enables him to see views of the solid from several angles and thus helps him generate a proper mental representation of the solid.

Sometimes rotating a solid is sufficient to check whether a certain conjecture is true or not, but often it is not easy to bring the solid into the desired position by using the three axes as axes of rotation. Therefore, a special rotation was added which rotates the solid so that a particular plane becomes the screen plane. After such a rotation the student can either view the entire solid projected on the screen plane or the screen plane alone. This special rotation has turned out to be one of the most powerful features of the software. It can be used to check whether a particular point or line lies in a plane, whether two given lines intersect, and so forth (Example 1).

Replay. An additional feature of the software allows the student to save the sequence of operations used in a construction and to replay it either on the same or on a different solid. This is particularly useful in order to find what remains invariant and what changes in going from a special case to a more general one.

The replay feature also allows teachers to prepare a sequence of operations (for instance, for teaching the concept of angle between a line and a plane) and then use the computer as an electronic blackboard in class, just replaying the prerecorded constructions.

From the Classroom

STEREOMETRIX is a tool; there is no curriculum firmly associated with it. It can be used in the classroom in different ways. For instance, it can be a tool for a student working on a problem, or an electronic blackboard for the teacher. Here we present four typical examples of classroom work. These examples do not constitute a curriculum. They are chosen to illustrate particular aspects of student activity with the software. Although they are ordered from easy to difficult, the question in Example 1, for instance, should only be asked after similar question for simpler solids (prisms, square, pyramid) have been solved.

Example 1: Picturing three-dimensional solids
Consider a right pyramid with vertices *A, B, C,* and *S* whose base *ABC* is a right triangle: $\angle BCA = 90°$. (A pyramid is called *right* if $SA = SB = SC$).
i)   Draw (on paper) the projection of the pyramid onto the plane of its base *ABC*; check your figure using the special rotation and explain the resulting figure.
ii) Draw the projection onto the plane *SAB*; check and explain.

iii) Draw the projection onto the plane *SAC*; check and explain. Where is the right angle?

Before approaching questions of this type, students are asked to rotate the solid and get an impression how it looks from various angles.

Discussion: In most views of the pyramid in this question it does not appear clearly that the base triangle is a right triangle. Thus it is difficult to imagine how the projections on the various faces will look. In fact, in order to answer correctly one needs to use definitions and properties; visualization alone is insufficient. Many students expect the projection onto *ABC* to look like Figure 2(a) and are surprised to find out that the correct answer is like Figure 2(b). In other words, their intuition or their previous experience led them to make the most frequent mistake in solving such problems without the software. However, work with the software immediately leads to a conflict between the expected and the obtained results, and this contradiction requires an explanation by means of analytic arguments. Thus because *SA* = *SB* = *SC*, *OA* = *OB* = *OC* where *O* is the projection of *S* onto *ABC*. Thus *O* is the center of the circumscribed circle of *ABC* and hence the midpoint of *AB*. By the way, younger students can deal with the same question on a purely visual level without entering into the analytic argument.

Figure 2(a)   *Figure 2(b)*   *Figure 2(c)*

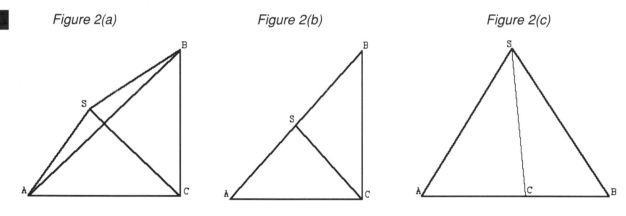

The answer to (ii) is even harder to guess (see Figure 2(c)). But the learning process is the same: the student draws what he expects, verifies, and then needs to explain by analytic argumentation. One way of doing this is to reason that the plane *SAB* is perpendicular to the plane *ABC*; thus the entire plane *ABC* is projected onto the line *AB*, and with it the point *C*. It is worthwhile to stress that the projection of *C* is not the midpoint of *AB* (unless *AC* = *BC*). The aim of (iii) is somewhat different. It is one of the cases contributing to the experience that the projection of a right angle does not usually appear to be 90°.

Figure 3

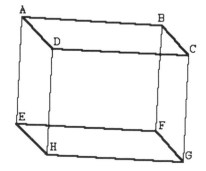

Example 2: Concept formation

Consider a straight prism whose base is a parallelogram (In a "straight" prism the "walls" are perpendicular to the bases.) Find the angle between the longest diagonal of the prism and one of its faces. To be more specific, let the base be *EFGH* with ∠*EFG* > ∠*FGH* (see Figure 3); construct and mark the angle between the diagonal *AG* and the face *BCGF*. Hints: Remember that in order to do this you need to find the foot of the perpendicular from *A* to the plane *BCG*. Start by conjecturing whether this foot is at *B*, at *C*, on the segment *BC*, on the line *BC* but not between *B* and *C*, or elsewhere in the plane.

Discussion: One of the first solids students get to know is the (rectangular) box. When learning to draw a two-dimensional projection of a box, students learn that (at least) four of the six faces appear as parallelograms. At a more advanced stage, students need to learn that the converse is not necessarily true: the picture in Figure 3 does not necessarily represent a rectangular box. Thus in the two-dimensional picture, there is no visual distinction between the prism in Figure 3 and a rectangular box.

When solving problems like this, students tend to relate to the picture as representing a rectangular box and ignore the given data. Thus they tend to think that
*   *AB* is the perpendicular from *A* to *BCGF*, and thus
*   ∠*AGB* is the angle between the diagonal *AG* and the face *BCGF*.
While intuition in this case is based on previous learning, it is incomplete and leads to errors.

Figure 4

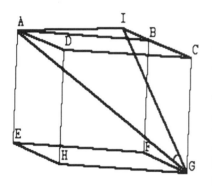

This error has been observed repeatedly when students solve similar problems with paper and pencil and when they make conjectures before checking with the software. When they use STEREOMETRIX to solve the problem, a conflict arises between what they expect and what they observe on the screen: the foot of the perpendicular is not *B* but a point beyond *B* on the extension of *CB*. The propagation of the error is thus prevented. More importantly, the students are given an opportunity to explain why the foot of the perpendicular is not at *B*. To do this, they may want to turn the solid to get a better view of the plane *ABCD*. This enlarges the collection of projections they know and thus resolves the conflict. This leads them to construct the correct point *I*, find out where it is situated and obtain the required angle ∠*AGI* between the diagonal and the face (see Figure 4). On a deeper level, the student learns to use definitions and data, rather than only visual information from pictures, during problem solving.

Example 3: Solution of complex problems
Consider an isosceles triangle *ABC* (*AB* = *AC*) and a line *k* through *A* in the plane bisecting *BC*. Choose a point *S* on *k* and construct the foot *D* of the perpendicular from *A* to *SBC*. What is the locus of points *D* as *S* moves on *k*?
Hints: Consider the pyramid *SABC*. The edge *SA* is on the line *k*. As *S* moves on *k*, find out what happens to the pyramid, to the perpendicular from *A* to *SBC*, and to its foot *D*. In particular, ask yourself whether all perpendiculars are in one plane. Once you have constructed several perpendiculars, make a conjecture about the locus and try to prove it.

Discussion: The leading idea from the didactic point of view is for the students to make conjectures, to check them, and then to prove them. The software can help in making conjectures and is very efficient in checking them. The development of a proof can be put off until later.

The first conjecture is suggested in the formulation of the problem: all foot points do indeed lie in one plane (otherwise it would be quite impossible, at least for high school students, to come up with any locus). After having constructed three foot points D_1, D_2, D_3, it is quite illustrative to turn their plane $D_1 D_2 D_3$ into the screen plane (using the special rotation) and to discover that not only do further foot points lie in the same plane but so do the points *A* and *S*.

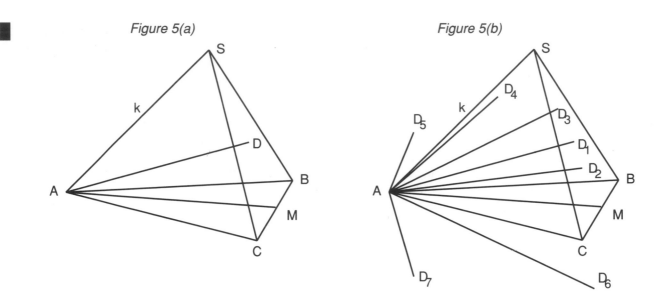

Figure 5(a)   *Figure 5(b)*

It is more difficult to come up with a conjecture for the locus. In fact, students have been observed to make all their choices for *S* in a rather restricted neighborhood of each other and thus not get a sufficiently wide distribution of the foot-points to make a conjecture. At this stage, it is useful to recommend that they make *S* approach *A*, on the one hand, and make it approach infinity, on the other hand. Here another didactic principle becomes important, namely the *visual-dynamic* approach. *S* moves on the line *k* and from the sequence of pictures the student is led to conclude how the pyramid and the perpendicular change as a consequence. Students may be asked further questions. Does the locus ever intersect *BC*? What happens if *SBC* is perpendicular to *ABC*? (The midpoint *M* of *BC* is the foot point.) Does *A* belong to the locus?

By now, students should be able to see clearly that (part of) a circle is generated in the plane *SAM*. At this stage, several more questions can be raised. Can you find the diameter of the circle on the screen? The circle seems to lack a part; can it be completed? By projecting on the plane *SAM*, it can be seen that the diameter is *AM*. The missing part of the circle can be built by choosing *S* on the other side of ("below") the plane *ABC*. It remains for them to construct a proof that the locus is the circle with diameter *AM*. Even at this stage, this computer-aided investigation can be useful for giving ideas on how to argue, and steps of the argument can be checked by using the software. For instance, one may ask about, conjecture, and check the mutual position of the planes *SAM* and *ABC*, concluding that the planes are perpendicular. Thus the perpendiculars from *A* to *SBC* are all in the plane *SAM*. Working in similar fashion, one may observe that in every case the foot-point *D* is on *SM* and thus *AD* is perpendicular to *SM*. In other words, the triangles *ADM* are right (see Figure 5; with the software, this can again be checked by special rotation). It follows that the locus of foot points is the locus of all points from which *AM* is seen under a right angle; that is, a circle with diameter *AM*.

Example 4: Projects

Let us now come back to the tetrahedron problem stated at the beginning of the paper and use it as an example of a project for able students:

Do the four altitudes of a tetrahedron meet in a single point?

This problem is nontrivial by most standards. In fact, Davis [1] cites Coolidge to the effect that most highly educated mathematicians do not know the answer. Although this remark implies that Coolidge probably knew the answer, we have not found it published. We thus set out to use STEREOMETRIX to help us with the problem. Note that this is not a "prove that ..." situation since we, the teachers, honestly did not know the answer. It is thus a situation where you first have to come up with a conjecture. It is obvious that there are tetrahedra for which the altitudes intersect (e.g., the regular tetrahedron); it also becomes clear rather quickly that this is not true for all tetrahedra; a pyramid whose base is a right triangle is an example. The question can thus be reformulated as

For which class of tetrahedra do the four altitudes meet in a single point?

One person whom we observed investigating the problem first reformulated it in this way and then used STEREOMETRIX to look at what happens for a pyramid whose base is an equilateral triangle *ABC*. Using the special rotation she found that in this case the altitudes do meet. She then moved the vertex *S* in a plane parallel to *ABC* and observed that this destroyed the coincidence. (The replay feature is useful here for repeating the construction on different tetrahedra.) More judicious experimentation made her realize that the altitudes were still intersecting in pairs if *S* was moved in such a way that the pyramid conserved a plane symmetry. At that stage she had amassed enough experimental information to ask herself about the general condition for two altitudes to meet. In fact, they meet if and only if two nonintersecting edges of the tetrahedron are perpendicular. From here through several more stages, some with and some without the software, she arrived at the conjecture that the four altitudes are coincident if (and only if) the projection of any vertex onto the opposite face is the intersection point of the altitudes of that face. She was able to prove this without direct use of the software but she indirectly exploited much of the experience she had collected during the "experimental" phase.

Complex problems like this are only suitable for gifted students. They should have the opportunity to work on such a problem during several sessions, with time for thinking matters over in the meantime. Such activity is likely to give them a feeling for the way research mathematicians work, a feeling for the fact that mathematics has many aspects of an experimental science.

Conclusion

The examples above demonstrate how four different goals can be achieved:
* improvement of students' spatial visualization by dynamic transition between two and three dimensions (Example 1),
* concept formation by means of direct use of definitions rather than only intuition (Example 2),
* solution of computational problems which involve conjecturing and checking (Example 3), and
* investigations based on experimentation, conjecturing and checking (Example 4).

We have shown how the integration of STEREOMETRIX in the teaching process can support these goals by helping teacher and student with drawing, conjecturing, checking, and visualizing.

The learning process is supported by the generation of conflicts between the student's intuition and the exact construction that appears on the screen; such conflicts can be resolved by analytic arguments. In summary, it was shown that the integration of STEREOMETRIX in the learning process helps the student to overcome learning difficulties that are characteristic of spatial geometry; it also helps the teacher to teach concepts, to use a teaching style based on conjecture and experimentation, and to encourage the most able students to carry out investigations of complex problems.

References

[1] Davis, Philip J., "Are there coincidences in mathematics?" *American Mathematical Monthly*, 88(1981), 311-320.

[2] Grenoble University, *Cabri Géomètre* (software). Laboratoire de structures discrètes et de didactique (IMAG), Grenoble, France, 1989.

[3] Hershkowitz, Rina, "Some psychological aspects of learning geometry," in P. Nesher and J. Kilpatrick, eds., *Mathematics and Cognition: A Research Synthesis by the International Group for the Psychology of Mathematics Education*. Cambridge University Press, in press.

[4] Hoyles, Celia, "Geometry and the Computer Environment," in J. Bergeron, N. Herscovics and C. Kieran, eds., *Proceedings*, Eleventh International Conference on the Psychology of Mathematics Education, Vol. II, 1987, 60-66. Montreal, Canada.

[5] Kramer, Emmanuel, Nurit Hadas, and Rina Hershkowitz, "Geometrical constructions and the micro-computer," in L. Burton and C. Hoyles (eds.), *Proceedings*, Tenth International Conference on the Psychology of Mathematics Education, 1986, 105-110. University of London Institute of Education.

[6] Osta, Imam, "L'outil informatique et l'enseignement de la geometrie dans l'espace," in J. Bergeron, N. Herscovics and C. Kieran (eds.), *Proceedings*, Eleventh International Conference on the Psychology of Mathematics Education, Vol. II, 1987, 31-38. Montreal, Canada.

[7] Parzysz, Bernard, " 'Knowing' vs 'Seeing'. Problems of the Plane Representation of Space Geometry Figures," *Educational Studies in Mathematics*, 19(1988), 79-92.

[8] Schwartz, Judah and Michal Yerushalmi, *The Geometric Supposer* (software), Sunburst, Pleasantville, NY, 1985.

Visualization in Geometry: A Case Study of a Multimedia Mathematics Education Project

Eugene A. Klotz

> ...a mathematician does not attain an understanding of a proof merely by checking that all the individual steps have been strung together according to the rules...What is crucial is to see through the technicalities to grasp the underlying ideas and intuitions, which often can be expressed concisely and even pictorially (Chernoff [7]).

The original meaning of the Greek word δε'ιχνυμι (deiknumi, "to prove") was to make visible or to show. It has been claimed that in reaction to this earlier visual approach, Euclid deliberately avoided visual arguments ([18], pp.190-197). By the 16th century, some translations of Euclid were unabashedly visual (this is discussed further below). On the other hand, the Legendre version of Euclid which ruled part of the 18th and 19th centuries proudly claimed to have no diagrams. But by mid-19th century, the Byrne Euclid (also discussed at the end of this paper) had diagrams which were even color coded! Quite clearly the world of mathematics oscillates between periods in which visualization aids are viewed as important pedagogy and eras in which such aids are viewed as a detriment.

We have only recently emerged from an era in which the prevailing fashion was to decorate advanced texts and lectures in a visual style which might be described as "Bourbaki basic" (that is to say, nonexistent). Whether or not this was good for mathematics education, mathematics flourished. This oscillation of sentiment for and against aids to visualization, together with the existence of excellent although blind geometer/topologists, suggests that we would do well to avoid unexamined enthusiasm for supplying visual aids for everything in sight, and ask (and try to answer) serious questions. At what level(s) are such approaches relevant? For whom? What works and what doesn't? Is there more to the story than visualization by itself? (e.g., is a tactile component necessary? Perhaps it's "I hear and I forget, I see and I remember, I don't do and I don't really understand"). It is worth noting that, at this time, the case for visualization rests almost entirely on anecdotal evidence.

This paper will contribute more anecdotal evidence. I will discuss a problem in mathematics education and the work we are doing to address that problem. The need here is great, the materials we have produced have been very well received, but we are just not set up to study the questions raised above. Nonetheless, I believe the questions are of great importance and are very hard to address satisfactorily. I hope they are not swept aside by a lot of pretty pictures, computer images, videotapes, and computer programs.

The Problem

Visual geometry skills are essential to many areas including, but by no means limited to, several branches of mathematics. Many people use these skills, including "crystallographers, biochemists, surgeons, aviators, mechanical shovel operators, sculptors, choreographers, and architects" ([3], p. 4). Task forces and study groups have consistently found that our students are lacking in these abilities ([2], p. 3, 7, 18, 23; [13], p. 4; [10], p. 96; [1], p. 21; [4], pp. 341, 342). See also the article by Dreyfus and Hadas in this volume for further discussion of this problem (and for a very different approach to its solution).

Eugene Klotz is Professor of Mathematics at Swarthmore College and Director of the Visual Geometry Project.

Interesting tools have been on hand for some time for developing new educational approaches to the subject. Philip J. Davis argued in 1974 that "visual geometry ought to be restored to an honored position in mathematics. Computer graphics comprising animation and color offers the possibility of going far beyond conventional drawings" ([8], p.113). Unfortunately, these tools have rarely been used for educational purposes. As Grünbaum says, "It is a curious fact that the amount of visually stimulating material actually seen by our students seems to have remained unchanged or even decreased, even though the possibilities for presenting visual mathematics, and geometry in particular, have expanded beyond what could have been imagined even relatively recently" [12].

The Visual Geometry Project was formed in response to the need for three-dimensional visual materials in the schools. The target audience was the high-school geometry student, but our materials have proved valuable to a much wider population. The project is the idea of Professor Doris Schattschneider, a geometer at Moravian College, and the author, a lover of geometry and of technology.

From the outset, we wanted the project to be able to provide high-quality visual images, preferably computer-generated to take advantage of the possibilities of animation. These possibilities go beyond the real world, for example, in rendering the opaque transparent, and in passing one object through another. We also felt it necessary to have some sort of tactile manifestation of these images, to snap the students out of television-watching mode, and because there seems to be some consensus that these sorts of "manipulatives" are genuinely useful in learning. (For example, in studying the childhoods of eminent persons who exhibit exceptional visual-spatial creativity, Roger N. Shepard has identified the access to and subsequent fascination by tangible geometric objects as a common characteristic [16], p.155.)

Finally, we thought it worthwhile to develop some sort of interactive computer programs. It is an article of faith with me that this medium is one of great educational potential, and it seems in dire need of careful attention at this time if it is to achieve its potential. We will discuss this component later.

Three-Dimensional Visualization

Here is a brief checklist of available three-dimensional media, along with our reasons for accepting or rejecting them.

Books: Static two-dimensional images of three-dimensional figures are all too limited ([8]; common sense). Three-dimensional pop-up figures can be splendid, as used in the Dee Euclid of 1570 [9], for example. Unfortunately, such figures are hard to design and expensive to execute. Stereograms — drawings which appear three-dimensional when viewed through two-color glasses (see, for example [11] or [14]) — ought to be a very effective educational tool. Unfortunately, a lot of people don't like them, and a significant number of people are visually not able to get them to work. Computer-generated holograms offer great promise as convincing and useable flat representations of three-dimensional figures, but not for a reasonable price at this time.

Standard microcomputer images: As the article by Dreyfus and Hadas in this volume attests, commonly available microcomputers can be used for three-dimensional images — provided the images are sufficiently simple. Unfortunately, we felt the need for more complicated images. Various gadgets make it possible to turn the screen into a stereoscope or viewmaster "real" three-dimensional image, but I suspect that there would be poor acceptance from the anti-stereogram contingent. I'm sure there would be problems with non-standard gadgetry.

Graphics computers: Some "super microcomputers" now allow real-time manipulation of complicated and convincing three-dimensional images. Unfortunately, the price of such machines will likely keep them out of the classroom (and, alas, my lab as well) for the rest of this millennium.

Stored computer-generated images: The presentation medium can be standard videotape, laserdisk, "CDG's," etc. Only standard videotape is universally and standardly available. To successfully impart visual information to a generation of students brought up on computer-generated television advertising it is necessary that the graphics frame buffer be of reasonably high resolution, in terms of both pixels and color (for example, to remove jagged stairstep effects through anti-aliasing). Moreover, the videotape needs to be laid down frame-by-frame, rather than shot directly from the computer screen.

Models: As I've indicated, actual three-dimensional objects were a necessary part of our plans. However, model-making is usually a labor of love and rarely a commercially viable enterprise; it's hard to find good, cheap models. The best material one can hope for at a reasonable price for immediate mass usage is cardboard. (For example, for some time we've been fruitlessly looking for a supply of cheap clear plastic boxes which are approximately 3" cubes).

Because we felt the need to produce materials which could be immediately and widely used, our choices from the above list are pretty clear: computer-generated images stored on videotape, and cardboard models.

Educational realities have dictated accoutrements to go along with our media arsenal. Workbooks are needed to suggest activities for students in this unfamiliar terrain. Teacher versions of the workbooks are needed, with background and answers. Separate videotapes for teachers are needed showing the novel materials being used in classrooms, and providing additional background. Both the workbooks and the teacher videos have proved to be very valuable (and very time consuming to produce).

Our Setup

Our first task was to set up a computer animation system to produce broadcast-quality videotapes devoted to three-dimensional geometry. When we began, the commercial software available was prohibitively expensive for true 3-D graphics, so our plan was to write our own. (It also seemed that home-grown software would be more flexible and malleable than a standard turnkey system). We based our system on a substantial three-dimensional modelling environment, the Berkeley UniGrafix system by Carlo Séquin, et al, that runs under the UNIX operating system and is in the public domain. This was ported to our microcomputer environment, and we adapted it to include color.

Our equipment includes two PC-clones running Xenix (for memory management purposes) with special videographics boards, two 3/4" U-Matic videotape recorders capable of frame-by-frame animation, and a good quality video camera for introducing a bit of reality into our videotapes.

The cost of professional quality video equipment prohibits computer animation from becoming a cottage industry at this time, but it is now clear to us how powerful computer animation can be in studying geometrical objects, and we hope that other educators will choose to enter this field, since video production via computer animation is a labor-intensive, slow process and the educational community could use more workers. We will be happy to provide our software to interested educators, but we warn that it requires a very specialized environment.

Our Videos

For our first unit we produced a short videotape on a single geometric figure, the stella octangula, shown in Figure 1. (We will not attempt to show illustrations from our videotapes, since without motion they lose their effectiveness). It appears that the video and workbook materials (including manipulatives) are suitable for a surprisingly wide audience, from seventh graders to in- and pre-service teachers. This range was not expected, but it is certainly welcome. In part, it seems a tribute to our visual approach.

Figure 1: *The Stella Octangula (stellated octahedron)*

As an example of the geometrical insights obtained with the models and video, one of the goals is to compare relative volumes of the stella octangula and associated polyhedra (such as the cube which is its convex hull) without using formulas. This can be done just by using the theorem which states that two pyramids of the same height and base area have the same volume. Since many teachers and some older students wish to grab a formula first and think afterward, this seems to be a useful corrective.

We have recently completed a unit on the Platonic solids and are now working on a unit dealing with three-dimensional symmetry. After that, we will probably work on a unit on drawing three-dimensional figures.

The Geometer's Sketchpad

We felt the need to develop a program with which one could *interactively* explore the two-dimensional implications of our three-dimensional work. For example, with the videotape on symmetry we wish students to be able to conveniently work with the basic symmetry operations in the plane. For our projected unit on drawing three-dimensional figures, it will be helpful for students to have some means of creating perspective drawings so that they can examine the consequences of changing vanishing points, horizon lines, and the like (Figure 2).

Figure 2: *Different orientations of the same staircase with a common vanishing point.* (Note: Figures 2 through 8 are images as they appear on the computer screen in *The Geometer's Sketchpad*; the program will have an option for printing images more precisely in PostScript® format, as in Figure 1)

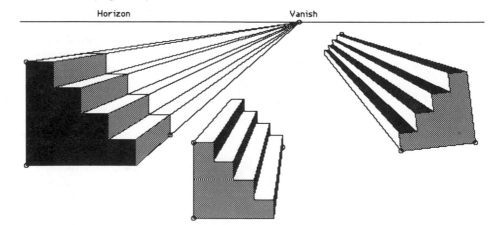

More generally, drawing figures is important for visualization, perhaps especially in two-dimensional geometry. There are many interesting areas where I've felt hampered as a teacher by the difficulty in making convenient but convincing drawings, either on the blackboard or for homework assignments, for example, even something as visually simple as Napoleon's theorem, Figure 3 (or, the nine-point circle, Morley's theorem, inversion in circles; think of anything that's fun). It is even more difficult to make a series of drawings which convinces one that a conjecture is true in all cases or from which one deduces additional facts (for example, in certain locus problems). Teachers and students have a need for a good, general purpose drawing program aimed at geometry. Our program, *The Geometer's Sketchpad,* allows the user to make accurate geometric drawings and measurements, to replicate a series of ruler-and-compass constructions on different figures, to build figures by specifying parts and properties, and to continuously (interactively) transform figures.

A computer program which purports to be interactive cries out for a useful and convenient user interface. We subscribe to that developed for the Xerox Star™, which was the basis for that of the Apple Macintosh™, so this became our target machine; we find its use of visual metaphors compelling. In addition, the "see and point" approach and emphasis on graphics design seem pertinent to geometry programs. It is interesting to note that the user interface together with subject matter can apparently influence the form of a program. We know of another excellent two-dimensional geometry program for the Macintosh being independently developed, *Cabri Géomètre* [6]. Our two programs, although quite different in many ways, also exhibit strong similarities. In addition to shaping programs the user interface seems to be contributing to new tools for visualization in plane geometry, as we report below.

Scripting

In 1985 a groundbreaking program by Judah Schwartz and Michal Yerushalmy appeared, *The Geometric Supposer,* which allowed the user to try out conjectures on different sets of related geometric objects [15]. *The Geometer's Sketchpad* generalizes this approach to broader sets of objects, and it not only remembers visual/tactile constructions, but also produces a written textual "transcript" of the constructions. This technique, an expansion of the computer science technique of "macro recording," is powerful in that it allows students to create an algorithm without typing, by having the program "watch" and store the student's actions.

Figure 3: *Two views of Napoleon's Theorem.* The centers *A', B', C'* of the equilateral triangles erected on the sides of any triangle *ABC* are the vertices of an equilateral triangle.

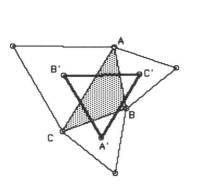
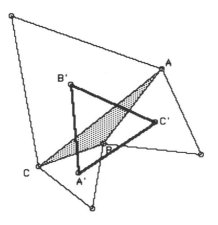

Figure 4: *A construction and associated script*

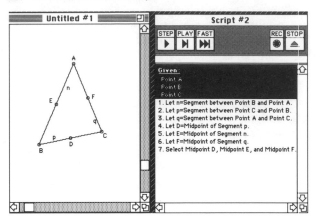

As an example, suppose we've asked the student to investigate the sequence of "midpoint triangles" of a given triangle (that is, take the triangle whose vertices are the midpoints of the given triangle, the triangle whose vertices are its midpoints, and so forth). The student might take points *A*, *B*, and *C*, construct the sides of triangle *ABC* and take the midpoints of these segments. The student can ask to record these operations as a script, as illustrated in Figure 4.

The computer generates a verbal description of the student's actions. The script can be applied to any figure which matches its "givens" (3 points); for example to the midpoints of the sides of triangle *ABC*, and to the midpoints of the triangle it would generate, and so on. It is apparent, after you've tried it, that this sequence closes in on a single point (the centroid of the original triangle) (Figure 5).

Scripting goes from the visual/manual to the verbal. It is our hope that it can aid those who have difficulty in visualizing written descriptions by making it easy for them to see step-by-step verbal descriptions of the figures they construct.

Dragging

One is tempted to apply the "see and point" approach to user interface as broadly as possible within a program. As an example, *The Geometer's Sketchpad* allows objects which are not otherwise constrained to be freely moved (by "grabbing with the mouse"), preserving the geometric relationships (such as a point being a midpoint) which apply. In general, this type of deformation of a figure can give the user a type of visualization, "visualization-by-perturbation", heretofore unavailable. With this, the user can see what a drawing would look like if small (or large) changes were made in the initial figures of a construction. This is a very powerful tool.

Figure 5: *Successive applications of the script in Figure 4*

To continue the above example, our student explorer has noticed that the midpoint triangles of the particular triangle form a nested sequence, which (apparently) converges to a single point. The student could deform the original triangle by dragging one of its vertices (in effect producing a projection

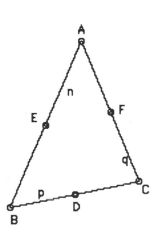
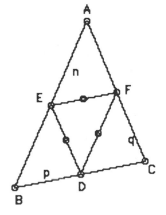
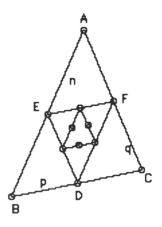

of the original configuration) and would become convinced that there was nothing special about the original triangle; one apparently has a nested sequence for any triangle (Figure 6).

Figure 6: *Two deformations of a sequence of midpoint triangles*

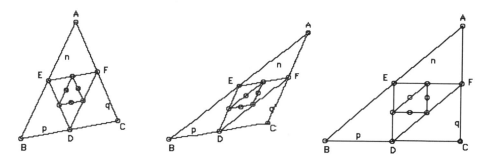

To determine the point of convergence, we might suggest that the student draw in the medians of the original triangle. After several deformations by dragging of vertices it would become clear that the point of convergence lies on all the medians, and hence the medians are concurrent (Figure 7).

It would be easy to justify the necessary steps suggested by this visualization and come up with a limiting argument which proved that the medians of a triangle are concurrent (and also gives the teacher a nice transparent situation in which to talk about limits).

When a figure is deformed by dragging, *The Geometer's Sketchpad* allows loci of objects to be traced out. That is to say, loci actually can be viewed as loci. A visualization concept, usually constrained to the mind's eye, has become manifest (Figure 8).

Visualization Aids in Geometry Education

There have been many ingenuous attempts to improve visualization in geometry, most of which have been ignored or forgotten. I've already mentioned the Dee Euclid of 1570 [9], which has pop-up figures in Book XI, together with clear and careful instructions for their use. As evidence that there is not much new under the sun, this first English Euclid also contains a number of interesting views of the Platonic solids which we show in our "groundbreaking" videotape, and it has the same type of nets for their construction as in our workbooks.

For those interested in the introduction of new visualization techniques, I must also point out the wonderful Byrne Euclid of 1847 [5], illustrated in Color Plate 2, "in which coloured diagrams and symbols are used instead of letters

Figure 7: *Different deformations with medians*

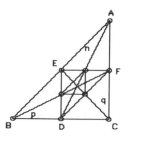
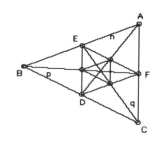
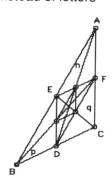

Figure 8: *The parabola as a locus as point P is dragged*

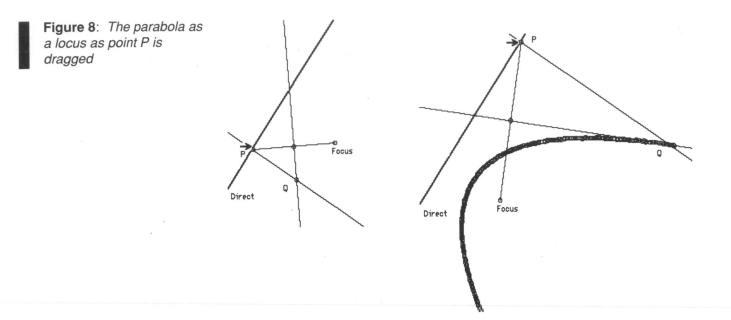

for the greater ease of learners," so that, for example, an angle would be referred to by a color, not a name (finding any occurrence of the angle becomes almost instantaneous). Like most of us who attempt to introduce new visualization techniques, Byrne had his anecdotal evidence (supplied by Horace), but unlike many of us, Byrne also had experimental evidence which he claimed showed that students could learn geometry his way in less than a third of the time necessary for traditional methods ([5], p. xii). Like some of us today, he linked his technique with the mathematics education of women (for him a recent educational innovation). Like almost all of us who attempt to introduce new techniques, Byrne was confident of their importance and wide applicability: "We shall for the present conclude by observing, as the senses of sight and hearing can be so forcibly and instantaneously addressed alike with one thousand as with one, the *million* might be taught geometry and other branches of mathematics with great ease, this would advance the purpose of education more than any thing that might be named, for it would teach the people how to think, and not what to think; it is in this particular the great error of education originates" ([5], p. xvii).

In common with all too many of us with new approaches to visualization, Byrne made a mistake; he didn't take adequately into account the high cost of his new technology (printing books with colored figures) for the education marketplace.

Closer to the present, G.F. Pearce's splendid 1977 stereogram approach to descriptive geometry [14], seems to have suffered the usual stereogram fate. Are all our wonderful new visualization technologies destined to be of only antiquarian interest?

I think that both computer-generated images (supplemented with some appropriate hands-on materials) and good interactive computer programs will prove not only worthwhile, but also will be accepted by the educational establishment. My reasons (anecdotal) have to do with seeing our own video/ model materials being used with a wide variety of clientele. Even modulo the salutary effect of novelty, there seems to be something there. One also obtains such an inter-ocular effect (it hits you right between the eyes) with both our two-dimensional geometry program and *Cabri Géomètre* — they clearly offer much more than simple drafting aids, more than concrete embodiments of the old mind's eye visualization.

There is one serious limitation to the broader development of the visualization media we have discussed: money. Constructing and storing high-quality computer-generated images is an expensive process and there are only a few receptive sources of adequate funds. The design and construction of computer programs may seem simple and cheap and possible by almost any mathematician. However, designing and constructing useful and *useable* programs is a much harder proposition. User interface problems are particularly thorny; there have been quite a few programs which only the programmer could use — or would want to. My own experience is that a good professional programmer is a necessity for my projects, even though I love to program. This ups the ante quite a bit.

Our mathematics education project has been very fortunate to work and experiment with a variety of media which seem particularly apt for visualization: experience with users leads us to believe that computer-generated images (with models) are important tools for visualization, and two-dimensional computer programs can make unanticipated educational contributions. It is unlikely that fresh media by themselves can stimulate the solution to the educational problems which face mathematics, but they can certainly help.

Those of us who believe that visualization has a more important role to play in mathematics education should be heartened by the possibilities available today, sobered by the history of visualization techniques, and willing to ask hard questions about this approach.

References

[1] *Academic Preparation for College,* The College Board, New York, 1983.

[2] *An Agenda for Action.* Reston, VA, National Council of Teachers of Mathematics, 1980.

[3] Baracs, Janos, "Geometrical Perception of Space," *Structural Topology,* 4(1980), 4.

[4] Brown, Catherine A., Thomas P. Carpenter, Vicky L. Kouba, Mary M. Lindquist, Edward A. Silver, and Jane O. Swafford, "Secondary School Results for the Fourth NAEP Mathematics Assessment: Algebra, Geometry, Mathematical Methods, and Attitudes," *Mathematics Teacher,* May 1988, 337-347.

[5] The Byrne Euclid was the work of Oliver Byrne, printed (very expensively) by Charles Whittingham and published by William Pickering in 1847. It is available in special collections and in antiquarian bookstores. I wish to thank the staff of F. Thomas Heller Rare Books of Swarthmore for both access to the book and much useful information. Thanks also to Special Collections, University of Delaware Library.

[6] *Cabri Géomètre* (software), Laborde, Jean-Marie et al, Laboratoire de Structures Discrète et de Didactique (IMAG), Grenoble, France, 1989.

[7] Chernoff, Paul, Letter in *Science,* 193(1976), 276.

[8] Davis, Philip J., "Visual Geometry, Computer Graphics, and Theorems of Perceived Type," *Proceeding of Symposia in Applied Mathematics American Mathematical Society,* 1974.

[9] The Dee Euclid, London, 1570, is available in special collections and in antiquarian bookstores. I wish to thank the staff of F. Thomas Heller Rare Books of Swarthmore, Pa. for both access to the book and much useful information.

[10] *Educating America for the 21st Century*, The National Science Board Commission on Precollege Education in Mathematics, Science, and Technology, 1983.

[11] Fejes Tóth, L., *Regular Figures,* Pergamon, New York, 1964.

[12] Grünbaum, Branko, "Shall We Show Them Some GEOMETRY?" 1984 lecture given at the "Visual Mathematics in the Undergraduate Curriculum" session, Mathematical Association of America summer meeting, August 16, 1984.

[13] *The Mathematical Sciences Curriculum K-12: What is still fundamental and what is not*, The Conference Board of the Mathematical Sciences, 1982.

[14] Pearce, G.F., *Engineering Graphics and Descriptive Geometry in 3-D,* Macmillan of Canada, 1977.

[15] Schwartz, Judah and Michal Yerushalmy, *The Geometric Supposer* (software), Sunburst, Pleasantville, NY, 1985.

[16] Shepard, Roger N., "Externalization of Mental Images" in Randhawa, Bikkar, and Coffman, William, *Visual Learning, Thinking, and Communication.* Academic Press, New York, 1978.

[17] "Study finds a Dropoff in Spatial Skills," Associated Press, *Philadelphia Inquirer*, 7/25/85, 19B.

[18] Szabó, Arpád, *The Beginnings of Greek Mathematics.* D. Reidel, Boston, 1978.

Acknowledgements

The work of the Visual Geometry Project described above received major support from the National Science Foundation Directorate for Science and Engineering Education, Division of Materials Development, Research, and Informal Science Education. *The Geometer's Sketchpad* program and the videotapes and workbooks are available from Key Curriculum Press, Berkeley, California.

Color Plates

Color Plate 1: *One of the George Francis/Donna Cox hyper-dimensional "Venus" homotopy figures. Copyright, Donna J. Cox. From the paper by Steve Cunningham, p. 67.*

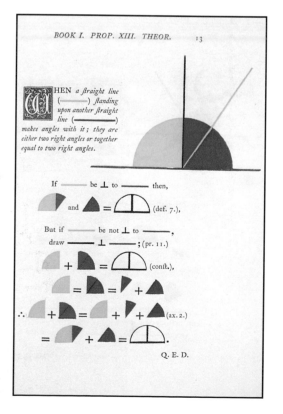

Color Plate 2: *A theorem from the Byrne Euclid, from Special Collections, the University of Delaware Library. Photograph by Photographic Services Department, University of Delaware. From the paper by Eugene A. Klotz, p. 95.*

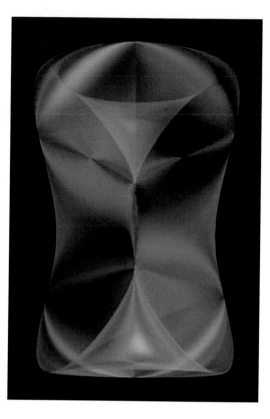

Color Plate 3: *A normal tube surface around a knotted space curve, from the paper by Thomas Banchoff, et al, p. 165.*

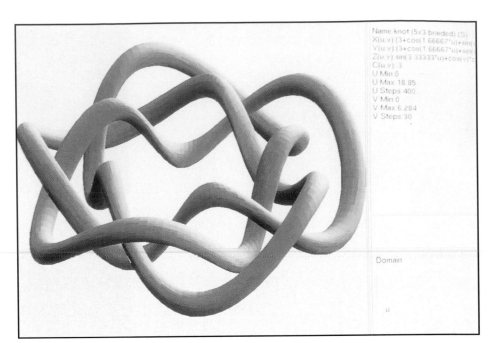

Color Plate 4. *Convergence diagram color coded by number of iterations for*
$f(x) = e^x + sin(x) + x^2$
    UL = Secant
    UR = Newton
    LL = Halley
    LR = Aitken
from the paper by Valerie Miller and G. Scott Owen, p. 197.

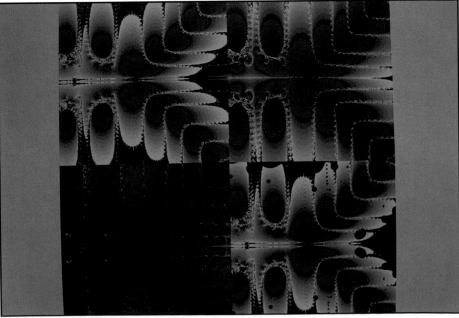

Color Plate 5. *Color coded by cost per iteration for*
$f(x) = e^x + \sin(x) + x^2$
    UL = Secant
    UR = Newton
    LL = Halley
    LR = Aitken
from the paper by Valerie Miller and G. Scott Owen, p. 197.

Color Plate 6. *Color coded by root and iteration for*
$f(x) = x^6 - 1$ *using Newton's method and* $|f(x^k)| < $ *tolerance from the paper by Valerie Miller and G. Scott Owen, p. 197.*

Color Plate 7: *Newton's method on* $x^7 - 1 = 0$
regions
UL = [-5,5] x [-5,5]
UR = [-5,-3.75] x [-0.625,0.625]
LL = [-5,-4.7] x [-0.15,0.15]
LR = [-4.8,-4.7] x [0.1,0.2]
from the paper by Valerie Miller and G. Scott Owen, p. 197.

Color plate 8: *Newton's method on $x^6 - 1 = 0$*
regions
UL = [-2,2] x [-2,2]
UR = [$\sqrt{3}/2-0.2$,$\sqrt{3}/2+0.2$] x [0.3,0.7]
LL = [$\sqrt{3}/2-.02$,$\sqrt{3}/2$] x [0.485,0.505]
LR = [$\sqrt{3}/2-0.005$,$\sqrt{3}/2-.002$] x [0.496,0.499]
from the paper by Valerie Miller and G. Scott Owen, p. 197.

Color Plate 9: *The Galton Board (above) and Toad's Ruin (below) from GASP, from the paper by David Griffeath, p. 215.*

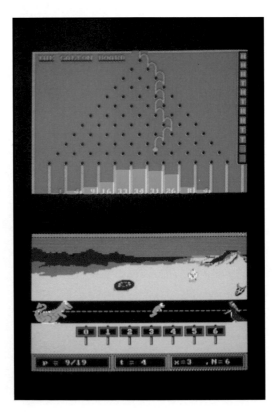

Intuition and Rigour:
The Role of Visualization in the Calculus

David Tall

Visual intuition in mathematics has served us both well and badly. It suggests theorems that lead to great leaps of insight in research, yet it also can lead up blind alleys of error that deceive. For two thousand years Euclidean geometry was held as the archetypal theory of logical deduction until it was found in the nineteenth century that implicit visual clues had insinuated themselves without logical foundation. An example is the implicit idea that the diagonals of a rhombus meet inside the figure, when the concept of "insideness" is not formally defined in the theory. Subtleties such as these caused even more pain in the calculus. So many fondly held implicit beliefs foundered when analysis was formalized. Comfortable feelings about continuous functions and the ubiquity of differentiable functions took a sharp jolt with the realization that most continuous functions are not differentiable anywhere. Once the real numbers had been axiomatized through the introduction of the completeness axiom, all intuition seemed to go out of the window. It is necessary to be so careful with the statement of theorems in formal analysis that any slight imprecision is almost bound to lead to falsehood. In such an atmosphere, visual mathematics has been relegated to a minor role, only that which can be proved by formal means being treated as real mathematics.

Yet to deny visualization is to deny the roots of many of our most profound mathematical ideas. In the early stages of development of the theory of functions, limits, continuity and the like, visualization was a fundamental source of ideas. To deny these ideas to students is to cut them off from the historical roots of the subject.

In this article we summarize research into visualization in the calculus over the last one and a half decades. It considers the strengths and weaknesses of visual imagery and relates this to the notions of intuition and rigour. It shows that visual ideas often considered intuitive by an experienced mathematician are not necessarily intuitive to an inexperienced student, yet apparently more complicated ideas can lead to powerful intuitions for the rigours of later mathematical proof. The theory of calculus is reconceptualized using the notion of "local straightness" — that a differentiable function is precisely one which "looks straight" when a tiny part of the graph is magnified. This gives a visual conception of the notion of a differential, which gives intuitive meaning to solutions of differential equations.

Research into mathematics education shows that students generally have very weak visualization skills in the calculus, which in turn leads to lack of meaning in the formalities of mathematical analysis. This paper, based on earlier work ([10] - [16]), suggests a way to use visual ideas to improve the situation.

The Value of Visualization

In mathematical research, proof is but the last stage of the process. Before there can be proof, there must be an idea of what theorems are worth proving,

David Tall is a Reader in Mathematical Education at Warwick University in the UK. After receiving his doctorate in mathematics at Oxford University with Michael Atiyah, he lectured in mathematics at the Universities of Sussex and Warwick before turning his attention to the psychology of mathematical thinking and obtaining a second doctorate in mathematics education with Richard Skemp. His particular interest is in the psychology of advanced mathematical thinking.

or what theorems might be true. This exploratory stage of mathematical thinking benefits from building up an overall picture of relationships and such a picture can benefit from a visualization. It is no accident that when we think we understand something we say "oh, I *see!*"

Figure 1: *curves with increasing x and y components*

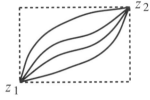

A good example is the famous Cauchy theorem in complex analysis that states that the integral of an analytic function around a closed curve which encloses no singularities is zero. When Cauchy stated an early version of this theorem, he thought of a complex number $z = x + iy$ analytically in terms of its real and imaginary parts. By analogy with the real case, he defined the contour integral

$$\int_{z_1}^{z_2} f(z)\ dz$$

between two points z_1 and z_2 along a curve whose real and imaginary parts are both either monotonic increasing or decreasing. As a formal generalization of the real case, this restriction on the type of curve is natural. But if we open our eyes and look at a picture, we see that such graphs (for x and y increasing) are a restricted set of curves lying in a rectangle with opposite corners at z_1 and z_2 (Figure 1). Cauchy had to visualize the situation for a more general curve in the complex plane to give his theorem in the form for a closed curve that we know today. If great mathematicians need to think visually, why do we keep such thinking processes from students?

The Weakness of Visualization

Visualization has its distinctive downside. The problem is that pictures can often suggest false theorems. For instance, it was believed in the nineteenth century that continuous functions have at most a finite number of points where they may be non-differentiable. The idea that a function could be continuous everywhere and differentiable nowhere was too strange to be contemplated.

Likewise, graphical methods were often used to prove analytic theorems. For instance, it was considered satisfactory to give a visual proof of the intermediate value theorem that a continuous function on an interval $[a,b]$ passed through all the values between $f(a)$ and $f(b)$. The curve was considered as a "continuous thread" so that if it is negative somewhere and positive somewhere else it must pass through zero somewhere in between (Figure 2).

Figure 2: *the intermediate value theorem*

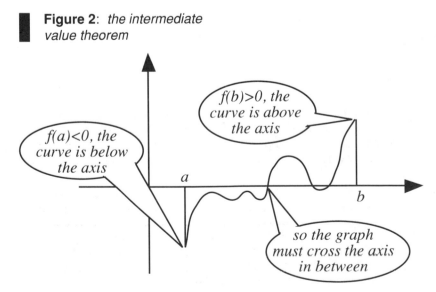

Yet we know that the function $f(x) = x^2 - 2$ defined only on the rational numbers is negative for $x = 1$, and positive for $x = 2$, but there is no *rational* number a for which $f(a) = 0$. Thus visualization skills appear to fail us. Life is hard.

But not *so* hard. What has happened is that the individual has inadequate experience with the concepts to provide appropriate intuitions. In this case a possible source of appropriate intuitions might be the numerical solution of equations on a computer where precise solutions are rarely found. Since most computer languages represent "real numbers" only as rational approximations, this may provide an intuitive foundation for the need to prove the intermediate value theorem rigorously.

Intuition and Rigour

A psychological view

In his essay "Towards a disciplined intuition," Bruner characterizes two alternative approaches to solving problems:

> In virtually any field of intellectual endeavour one may distinguish two approaches usually asserted to be different. One is intuitive, the other analytic ... in general intuition is less rigorous with respect to proof, more oriented to the whole problem than to particular parts, less verbalized with respect to justification, and based upon a confidence to operate with insufficient data. (Bruner [2], p. 99)

Some psychologists relate different modes of thinking to the two hemispheres of the brain. Glennon [6] summarizes the findings "from many research studies" (Figure 3).

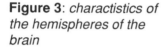

Figure 3: *charactistics of the hemispheres of the brain*

Left Hemisphere	Right Hemisphere
Verbal	Visuospatial
Gestural	(including
Logical	communication)
Analytic	Analogical, intuitive
Linear	Synthetic
Sequential	Gestalt, holist
	Simultaneous &
	Multiple processing
Conceptual similarity	Structural similarity

Other research shows that the breakdown is not always related in this precise physical way:

> Special talents ... can reside in the right brain or in the left. Clearly what is important is not so much where things are located, but that specific brain systems handle specific tasks. (Gazzaniga [5])

However, the principle underlying different modes of thinking remains, and we shall refer to these as the operations of the "metaphorical left and right brains," which may reside in these hemispheres in many individuals, but may be located elsewhere in others. The existence of different modes of thought suggests a distinction between intuitive thought processes and the logical thought demanded by formal mathematics. Intuition involves parallel processing quite distinct from the step-by-step sequential processing required in rigorous deduction. An intuition arrives whole in the mind and it may be difficult to separate its components into a logical deductive order. Indeed, it is known that visual information is processed simultaneously; only the result of this processing is made available to the conscious self, not the process by which the gestalt is formed (Bogen [1], Gazzaniga [4]). Taken to extremes, this suggests that the logic of mathematics may not be well served by an intuitive approach.

On the other hand, a purely logical view is also cognitively unsuitable for students:

> We have all been brainwashed by the undeserved respect given to Greek-type sequential logic. Almost automatically curriculum builders and teachers try to devise logical methods of instruction, assuming logical planning, ordering, and presentation of content matter ... They may have trouble conceiving alternative approaches that do not go step-by-step down a linear progression ... It can be stated flatly, however, that the human brain is not organised or designed for linear, one-path thought. (Hart [7], page 52)

> ... there is no concept, no fact in education, more directly subtle than this: the brain is by nature's design, an amazingly subtle and sensitive *pattern-detecting* apparatus. (*ibid.,* page 60)

> ... the brain was designed by evolution to deal with *natural complexity*, not neat "logical simplicities" ... (*ibid.,* page 76)

There is much evidence to show that the most powerful way to use the brain is to integrate both ways of processing: appealing to the (metaphorical) right brain to give global linkages and unifying patterns, whilst analysing the relationships and building up logical inferences between concepts with the left. This requires a new synthesis of mathematical knowledge that gives due weight to both ways of thought. In particular, it needs an approach which appeals to the intuition and yet can be given a rigorous formulation.

Geometric Concepts Need not be Intuitive

One of the reasons why the teaching of the calculus is in disarray is that concepts which expert mathematicians regard as intuitive are not "intuitive" to students. The reason is quite simple. Intuition is a global resonance in the brain and it depends on the cognitive structure of the individual, which in turn is dependent on the individual's previous experience. There is no reason at all to suppose that the novice will have the same intuitions as the expert, even when considering apparently simple visual insights. Mathematical education research shows that students' ideas of many concepts are not what might be expected. For example, because the formal idea of a limit proves difficult to comprehend in the initial stages of the calculus, it is usually introduced through visual ideas, such as the derivative being seen as the limit of a sequence of secants approaching a tangent.

Figure 4 : *secants and the limiting tangent*

Empirical research shows that the student has a number of conceptual difficulties to surmount. For instance, Orton [8] reported the following responses from 110 calculus students: when questioned what happens to the secants PQ on a sketched curve as the point Q_n tends towards P on the circle, 43 students seemed incapable, even when strongly prompted, to see that the process led to the tangent to the curve (Figure 4). There appeared to be considerable confusion in that the secant was ignored by many students, they appeared only to focus their attention on the chord PQ, despite the fact that the diagram and explanation were intended to try to insure that this did not happen. Typical unsatisfactory responses

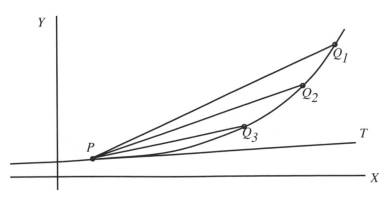

included: " the line gets shorter," " it becomes a point," and " the area gets smaller."

A similar question was asked in Tall [9]: in Figure 5, as $B \rightarrow A$ the line through AB tends to the tangent AT. True/False? Of a sample of nine 16-year-old students interviewed in depth (as part of a larger project), four said the statement was "true" but linked the symbol $B \rightarrow A$ to vector notation and visualized B as moving to A, along the line BA. For them the line (segment) BA certainly "tends" to the tangent, but in a completely unexpected sense. Meanwhile, another student considered the statement "false" because "way off at infinity the line AB and tangent AT would always be a long way apart no matter how close A and B become." Thus it is possible also to have an "incorrect" response for a very sensible reason.

Figure 5 : *a secant "tending" to a tangent*

As $B \rightarrow A$ the line through AB tends to the tangent AT.

True/False?

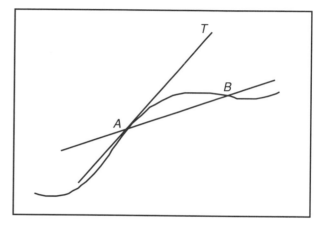

In Tall [11], a question to investigate the intuitive nature of the limiting process was given to 160 students about to start a calculus course in the UK, of whom 96 had already had some calculus experience (Figure 6).

Only 16 students (10%) obtained both the value $k + 1$ for the gradient (slope) of AB and 2 for the tangent, whilst 44 (24%) obtained $(k^2 - 1)/(k - 1)$ and 2. After the first two months of the calculus course, the numbers changed hardly at all to 17 (11%) and 38 (24%) respectively. Of these, only one student on the pre-test (who had already had calculus experience) and one student on the post-test allowed k to tend to one to find $k + 1$ tends to 2. On interviewing other students, it was clear that no limiting idea occurred to them. One who found the gradient of AB to equal $k + 1$ could see visually that the gradient of AT was about 2. It was suggested that as B got close to A, k would get close to 1 and $k + 1$ would get close to 2. I well remember the amazement on his face when he realized this for the first time.

Figure 6 : *is the limit concept intuitive?*

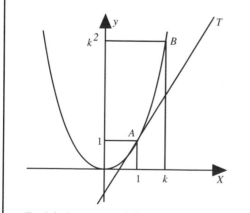

On the graph $y = x^2$, the point A is (1,1), the point B is (k, k^2) and T is a point on the tangent of the graph at A.

(i) Write down the gradient of the straight line through A, B ...

(ii) Write down the gradient of AT ...

Explain how you might find the gradient of AT from first principles.

In this experiment a spontaneous limit concept did not occur to *any* pupil with no calculus experience. This gives no support to the idea that the geometric limit is an intuitive concept. On the contrary, many other research investigations point to serious conceptual problems with the limit concept (Schwarzenberger and Tall [9], Cornu [3], Tall and Vinner [16], Sierpinska [17]). Students have difficulties because of the language, which suggests to them that a limit is "approached" but cannot be reached. They have difficulties with the unfinished nature of the concept, which gets close, but never seems to arrive. They have even more difficulties handling the quantifiers if the concept is defined formally.

A New Kind of Intuition for the Calculus

If we are to fill the gap in students' understanding of the calculus, then I hypothesise that we must find a way that is cognitively appealing to the student at the time the study commences yet has within it the seeds for understanding the formal subtleties that occur later. My analysis of the difficulty is that we will certainly not do this by making the concepts simpler. The alternative is to make them more complicated!

This is not as foolish as it sounds. The idea is to appeal to the visual patterning power of the metaphorical right brain in such a way that it lays down appropriate intutions to service the logical deductivity of the left.

The reason why nineteenth-century mathematicians found the concept of an everywhere continuous, nowhere differentiable function unintuitive was simply that they had not met a friendly example. Nor, I believe, have many current professional mathematicians. On one occasion I asked all the members of an internationally known mathematics department if they could furnish me with a simple proof of the existence of an everywhere continuous nowhere differentiable function. None of them could do this on the spot, though two could name a book where a proof could be found and one was even able to give the page number! I was equally unable to formulate such a proof at the time. If we professionals are so unable to give a meaningful explanation of a concept, what hope is there for our students? The answer lies in effective use of visualization to give intuition for formal proof.

A Locally Straight Approach to the Calculus

Given that the concept of limit seems such an unsatisfactory cognitive starting point for the study of calculus, and attempts at making it geometrically "intuitive" also fail, we need a subtly different approach. This is possible through an amazingly simple visual device. We know the gradient, or slope, of a straight line $y = mx + b$ is just the change in y-coordinate divided by the corresponding change in x-coordinate, but this fails for a curved graph. The answer is to magnify the picture. If a sufficiently tiny part of the graph is drawn highly magnified, then most of the familiar graphs look (locally) sraight.

Local Straightness

Drawing graphs accurately by hand is a major activity. But once students have some experience of drawing graphs, a graph-plotting program can be used to magnify the picture. This is best done with a function plotter with (at least) two graph windows: one for the original scale graph, the other to see a magnified smaller portion (Figure 7).

Given a little time to experiment, students will hypothesise that the more a graph is magnified, the less curved it gets. When it is suitably highly magnified, it will look locally straight. These students now have a significantly different mind set from traditional students. They are able to cast their eye along a graph and see its changing gradient. Their visual intuition is sharper.

Nonlocally Straight Graphs

Students just given simple examples of locally straight graphs are likely to be dangerously misguided. For just as nineteenth century mathematicians were convinced by their limited experience that "most" graphs are differentiable "almost everywhere," limited unguided experimentation can easily lead to the belief that all graphs are locally straight.

So we must make the experience more complicated immediately, before the mind is set. If asked to suggest graphs which are not locally straight, my experience is that students find it very difficult to make the first step. But once one or two examples are given, the floodgates open. It is now my preference in the first lesson to look at graphs like $y = |x|$ or $y = |\sin x|$, or $y = |x^2 - x|$ to see that they have "corners" which magnify to two half-lines with different gradients meeting at a point. It is also easy to jazz these up a little to add a tiny graph like $y = |\sin 100x|/100$ to a smooth graph to get, say $y = \sin x + |\sin 100x|/100$ which looks smooth, like $\sin x$, to a normal scale, yet has corners when magnified by a factor 100 or so.

Figure 7: *A locally straight curve (magnified a little)*

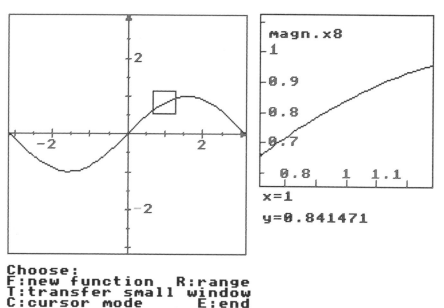

It is also interesting to focus on points where a graph may oscillate strangely, such as $f(x) = x \sin(1/x)$ at the origin (with $f(0) = 0$), or even $g(x) = (x + |x|)\sin(1/x)$. The latter is locally straight in one direction at the origin, but oscillates wildly in the other.

Nowhere Differentiable Functions

Even exhibiting curves with corners gives inadequate intuition. We must take our courage in both hands and go the whole way. The *Graphic Calculus* software (Tall [11], Tall, *et al*, [15]) includes a model of an everywhere continuous, nowhere differentiable function called the *blancmange function* after a custard pudding with a similar shape. The function is simply so wrinkled that wherever it is magnified it still looks wrinkled (Figure 8).

Figure 8: *A highly wrinkled function that is nowhere locally straight*

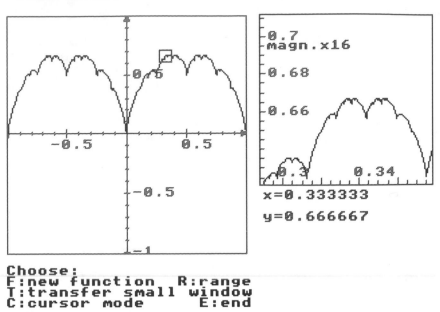

f(x)=bl(x)

Choose:
F:new function   R:range
T:transfer small window
C:cursor mode   E:end

The power of this function, and the recursive way that it is defined, is that it is easy to give an intuitive proof as to why it is nowhere differentiable. The argument (given in Tall [10]) is like this: The pictures in Figure 9 show saw-teeth $s_0(x)$, $s_1(x)$, $s_2(x)$ ... where successive ones are half the scale of the previous graph. The blancmange function is $bl(x) = s_0(x) + s_1(x) + s_2(x) + \cdots$ Mathematically this is a subtle definition, but cognitively it makes sense when the construction is seen in dynamic computer graphics building up partial sums, because the teeth soon get so small that, to a given degree of accuracy, they add nothing significant to the sum, which then does not visibly change.

Figure 9: *Constructing the blancmange function*

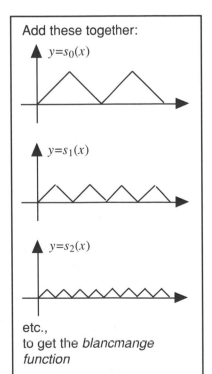

Add these together:

$y=s_0(x)$

$y=s_1(x)$

$y=s_2(x)$

etc.,
to get the *blancmange function*

If the first sawtooth is omitted and the second, third, fourth etc. are added, the resulting graph is clearly a half-size blancmange. So the blancmange is the first saw-tooth plus a half-size blancmange. Similarly, the sum can be broken down into the sum of the first two saw-teeth, plus the sum of the third, fourth, fifth etc. The sum of the first two is the second approximation to the blancmange, the sum of the remainder is a quarter-size blancmange. In general the blancmange can be seen as being the n th approximation with a $1/2^n$ size blancmange added. This is why it is so wrinkled. It has blancmanges growing everywhere!

Once it is understood that an everywhere wrinkled graph $y=bl(x)$ can be constructed, it is easy to see that the graph $n(x) = bl(1000x)/1000$ is a very tiny wrinkle indeed (it is smaller than 1/1000, and only shows up under magnification of 100 or so). In fact $y = \sin x$ looks just like $y = \sin x + n(x)$ but the first is locally straight and the second looks wrinkled under high magnification.

Thus, by visualizing we have broken the fetters of visualization. We can envisage two graphs which look the exactly the same at normal magnification, one differentiable everywhere, one differentiable nowhere.

Interactive Gradients

Once the idea of local straightness is established, all the standard derivatives can be conjectured by looking at the gradient (slope) of the graphs. In addition to using the computer, it is possible to carry the action out using a simple tool, designed by the School Mathematics Project in England and called the *gradient measurer*. This is a circular piece of transparent plastic with a diameter marked, which can turn around its centre. It is affixed to another transparent piece of plastic (by slotting into lips which overlap the diameter) on which is marked a vertical ruler in units, one unit horizontally away from the centre.

Figure 10 : *A tool for measuring gradients*

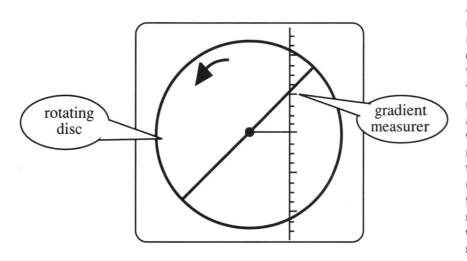

A student can place the gradient measurer over a point on the graph, rotate the disc until the marked diameter is visually in the direction of the graph at that point, then read off the gradient. Thus she or he can move the measurer along the graph and follow the changing gradient, as well as obtain approximate numerical results. (Presupposing, of course, that the graph is a faithful representation of the gradient, both in terms of having the same *x*- and *y*-scale and also not having any tiny wrinkles that cannot be seen at this scale.)

Computer Generated Gradients

It is an easy matter to program the computer to draw the numerical gradient $(f(x + c) - f(x))/c$ for a fixed, but small, value of *c*. This can be done in such a way that the student can superimpose a conjectured graph to compare with it. In this way it is possible to conjecture what the derivative of simple functions will look like. All the standard functions can be investigated in this way. By using this approach, the seed is sown that the gradient changes as a function of *x*.

Figure 11 : *Visualizing the global gradient function*

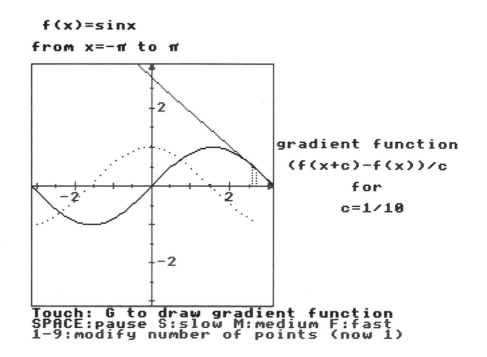

```
f(x)=sinx
from x=-π to π
```

gradient function
$(f(x+c)-f(x))/c$
for
c=1/10

```
Touch: G to draw gradient function
SPACE:pause S:slow M:medium F:fast
1-9:modify number of points (now 1)
```

Differential Equations: Undoing Local Straightness

Sketching solutions of a differential equation $dy/dx = f(x,y)$ by hand is a complicated process. But the tasks may be shared with a computer. The *solution sketcher* (Tall 1989) does this by showing a short line segment centred at a point (x,y) with gradient $f(x,y)$ (Figure 12).

Figure 12: *The solution sketcher*

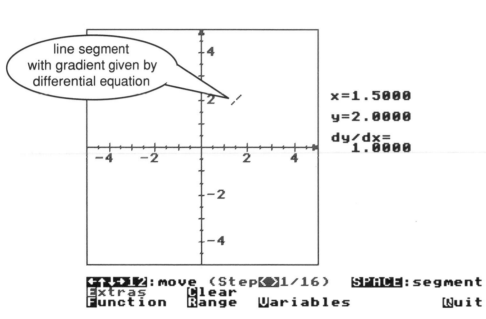

dy/dx=0.5y

line segment with gradient given by differential equation

x=1.5000
y=2.0000
dy/dx=
1.0000

◀▶▲▼:move (Step 1/16)   SPACE:segment
Extras   Clear
Function   Range   Variables   Quit

The line segment may be moved around, either using the mouse or cursor keys. As this happens the segment takes up the direction given by the differential equation $dy/dx = f(x,y)$. Clicking the mouse or touching the SPACE bar leaves a copy of the segment at the current position, allowing the user to build up a solution by following the direction specified in an *enactive way*. By this I mean that the user carries out the physical act of following a solution curve, whilst the computer calculates the gradient.

All the usual formal theory about the existence and uniqueness of solutions arises through enacting the solution process physically, providing powerful intuitions. There is a unique solution through every point (x,y) and it continues as long as the equation continues to specify the required direction.

Figure 13 : *Constructing a solution of a differential equation*

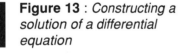
dy/dx=0.5y

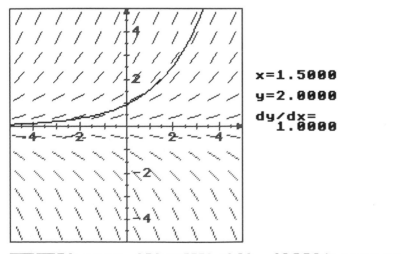

x=1.5000
y=2.0000
dy/dx=
1.0000

◀▶▲▼:move (Step 1/16)   SPACE:segment
Direction field Stream-line   Clear
Function Range Variables   Grid Quit

Figure 13 shows a solution constructed in this way. It is superimposed on a whole array of line segments whose gradients are given by the differential equation, showing the *global* trends of other possible solutions.

Solving differential equations such as $dy/dx = bl(x)$ where the right-hand side is continuous, but not differentiable, gives a solution curve $y = F(x)$ whose gradient satisfies $F'(x) = bl(x)$. Thus the function F is differentiable once, but not twice. Repeating the process several times can give a function which is differentiable n times but not $n + 1$, enabling the mental imagery to be developed to encompass such functions.

Integration

Integration is the idea of "cumulative growth," which is usually seen as calculating the area "under" a curve. This can be performed using thin rectangular strips, or methods such as the trapezium or Simpson rule. Visually, by taking a larger number of strips it becomes apparent that sum of

Figure 14: *Negative step and negative ordinate gives a positive area calculation*

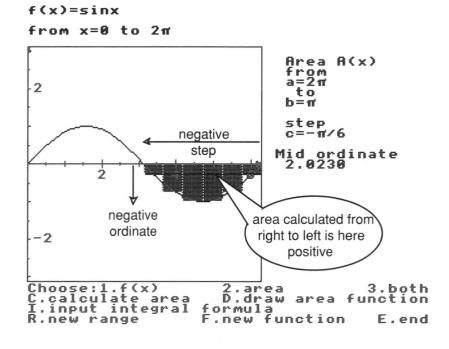

```
f(x)=sinx
from x=0 to 2π
```

Area A(x)
from
a=2π
to
b=π

step
c=-π/6

Mid ordinate
2.0230

negative step

negative ordinate

area calculated from right to left is here positive

```
Choose:1.f(x)        2.area           3.both
C.calculate area     D.draw area function
I.input integral formula
R.new range          F.new function    E.end
```

strip-areas is likely to give the value of the area under the curve. However, flexible software may be used for all kinds of investigations to give more powerful intuitions. For instance, most students and not a few teachers believe that the area is "positive above, negative below," but if the sign of the step is negative, the reverse is true. Figure 14 has a negative step and a negative ordinate in the range working backwards from 2π to π, giving a positive result when the curve is below the x-axis. Although this seems more complex than just giving a simple rule, it easily provides a complete mental picture of the four possible combinations of sign of step-direction and ordinate, with the dynamic movement giving powerful intuitions linking with signed arithmetic.

Figure 15: *The area function from x = 0 in both directions*

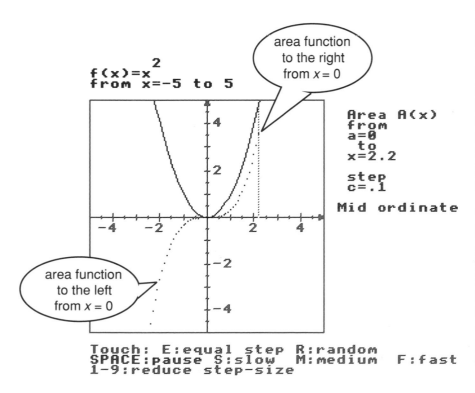

```
f(x)=x²
from x=-5 to 5
```

area function to the right from x = 0

area function to the left from x = 0

Area A(x)
from
a=0
to
x=2.2

step
c=.1

Mid ordinate

```
Touch: E:equal step R:random
SPACE:pause S:slow M:medium F:fast
1-9:reduce step-size
```

The area-so-far graph may be drawn by plotting the value of the cumulative area calculations from a fixed point to a variable point. Figure 15 shows the superimposition of two area calculations, first from $x = 0$ to $x = -5$ with the negative step -0.1, then from $x = 0$ to the right with positive step 0.1. Notice the cubic shape of the dots of the area curve, which experts will recognize as $y = x^3/3$. But we rarely consider this for negative x.

Undoing Integration: The Fundamental Theorem

Numerical Gradients and Areas as Functions

Many programs will draw numerical gradients (slopes) or numerical areas, but these are usually just a sequence of points plotted on the screen. An area graph plotted as in Figure 15 simply records the cumulative area calculations pictorially and does not remember them in any way. But the area under a graph of a given function, from $x = a$ to $x = b$, with a given step, say h, is calculated by a straightforward computer procedure. Given a suitably fast processor this can be calculated almost instantaneously and may be considered as a function depending on the graph and the values of a, b, and h.

■ **Figure 16**: *The area function for the blancmange and its derivative*

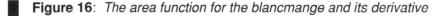
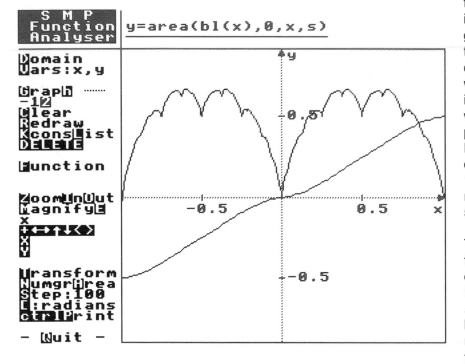

In the Function Analyser (Tall [14]), the expression *area*(*expr, a, b, h*) is interpreted as the area under the graph given by the expression *expr*, from $x = a$ to $x = b$ using the mid-ordinate approximation with step-width h. Thus *area*(sin x, 0, x, 0.1) is the area under $y = \sin x$ from 0 to x with step-width 0.1. Such is the power of the Archimedes computer in British schools that this graph can be drawn in less than three seconds with 100 intermediate points each requiring an area calculation for up to 50 strips.

The numerical area is now truly a function, which may be numerically differentiated like any other function. Figure 16 shows the area function *area*(*bl* (*x*), 0, *x*, *s*) under the blancmange function from 0 to x using strip-width $s = 0.05$. (The area function is the rather bland looking increasing function, not the pudding-like blancmange). Of course, this graph is not the exact area function, but it is a good-looking approximation to it, as good as one could hope to get on a computer screen. Notice that it looks relatively smooth, which it is if one ignores the pixellation problem, because the derivative of the exact area is the blancmange function. The exact area function for the blancmange is a function which is differentiable everywhere once and nowhere twice!

The other graph in Figure 16, by the way, may look like the blancmange function, but it is actually the graph of *area*(*bl* (*x*), *x*, *x+w*, *h*)/*w* (the numerical derivative of the numerical area function for the blancmange). Even though the blancmange function is nowhere differentiable, its area function is quite smooth and differentiable everywhere precisely once.

Stretching the Imagination for the Fundamental Theorem of the Calculus

One way to visualize the fundamental theorem is to imagine a tiny part of the graph stretched horizontally (Figure 17). In many cases the graph of a function stretches out to look flat; the more it is stretched, the flatter it gets. This is easy to see with a graph drawing program using a thin x-range and a normal y-range to stretch the graph horizontally in a standard graph window.

Figure 18 shows that if the area from a fixed point a to a variable point x is $A(x)$, then the area from a to $x + h$ is $A(x + h)$, so the change in area from x to $x + h$ is $A(x + h) - A(x)$. If the strip from x to $x + h$ is approximately a rectangle with width h and height $f(x)$, then its area is $A(x + h) - A(x) \approx f(x)h$ whence $(A(x + h) - A(x))/h \approx f(x)$. As h gets smaller, the graph gets pulled flatter and the approximation gets better, giving intuitive foundation for the fundamental theorem of the calculus: $A'(x) = f(x)$.

Figure 17: *stretching a graph horizontally*

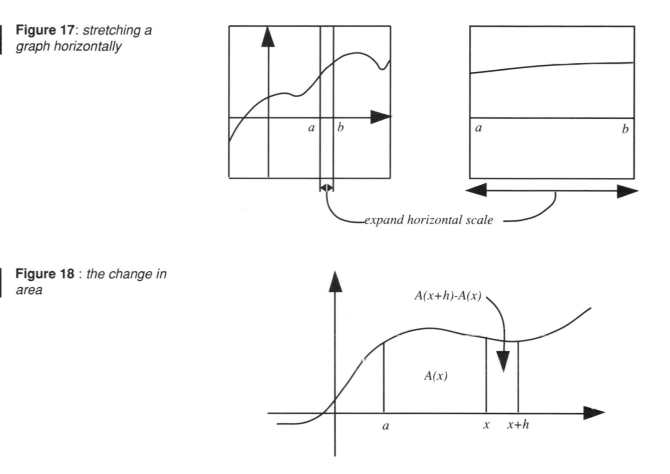

expand horizontal scale

Figure 18 : *the change in area*

From Intuition to Rigour

A formal proof of the fundamental theorem requires the notion of continuity. This notion is usually confused intuitively by students, teachers, and mathematicians alike. Ask anyone with some knowledge of the concept to explain what it means and the likely answer is that a continuous function is one whose graph has "no gaps;" its graph can be drawn "without taking the pencil off the paper," and the like (Tall and Vinner [16]). These ideas are not the intuitive beginnings of continuity but of "connectedness" which is mathematically linked, but technically quite different.

Continuity can be seen to arise from the horizontal stretching of graphs in the fundamental theorem. Consider a simplified model of what is happening in stretching the graph to confine it within a horizontal line of pixels. Suppose that graph picture has middle x-value $x = x_0$ and the point $(x_0, f(x_0))$ on the graph is in the middle of a pixel whose upper and lower values are $f(x_0) - e$

and $f(x_0) + e$. To fit the graph in a horizontal line of pixels means finding a small x-range from $x_0 - d$ to $x_0 + d$ so that for any x in this range the value of $f(x)$ lies in the "pixel range" between $f(x_0) - e$ and $f(x_0) + e$ (Figure 19). This gives the formal definition of continuity:

> The function f is continuous at x_0 if, given any specified error $e > 0$, there can be found a (small) distance d such that whenever x is between $x_0 - d$ and $x_0 + d$, then $f(x)$ is between $f(x_0) - e$ and $f(x_0) + e$.

The ramifications of this definition take months, even years, to understand in full, but it has an appealing intuitive foundation: a continuous function is one whose graph has the property that any suitably tiny portion stretched horizontally will pull out flat.

Figure 19 : *The concept of continuity through horizontal stretching*

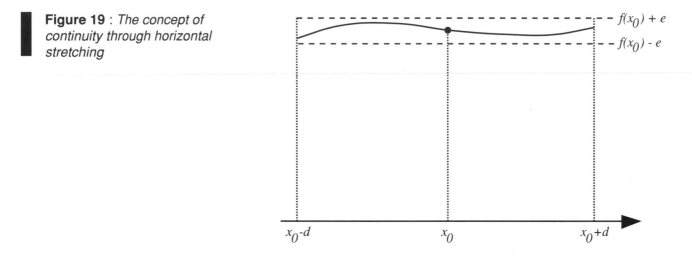

Conclusions

By introducing suitably complicated visualizations of mathematical ideas it is possible to give a much broader picture of the ways in which concepts may be realized, thus giving much more powerful intuitions than in a traditional approach. It is possible to design interactive software to allow students to explore mathematical ideas with the dual role of being both immediately appealing to students and also providing foundational concepts on which the ideas can be built. By exploring examples which work and examples which fail, it is possible for students to gain the visual intuitions necessary to provide powerful formal insights. Thus intuition and rigour need not be at odds with each other. By providing a suitably powerful context, intuition naturally leads into the rigour of mathematical proof.

References

[1]   Bogen, J. E., "The Other Side of the Brain: 2. An Appositional Mind," *Bulletin of the Los Angeles Neurological Society*, 34(1969), 135-162.
[2]   Bruner, J. S., *Relevance of Education*, 98-113, Oxford University Press, 1974.
[3]   Cornu, B., "Apprentissage de la notion de limite: modèles spontanés et modèles propres," *Actes due Cinquième Colloque du Groupe International P.M.E.*, Grenoble (1981), 322-326.
[4]   Gazzaniga, M. S., "Cerebral Dominance Viewed as a Decision System," *Hemisphere Function in the Human Brain*, Dimond, S. and J. Beaumont (eds), Elek, London, 1974.

[5] Gazzanigna, M. S., *The Social Brain: Discovering Networks of the Mind*, Basic Books, 1985.

[6] Glennon, V. J. "Neuropsychology and the Instructional Psychology of Mathematics," *Proceedings*, Seventh Annual Conference of the Research Council for Diagnostic and Prescriptive Mathematics, 1980, Vancouver B. C., Canada.

[7] Hart, L. A., *Human Brain and Human Learning*, Longman, New York, 1974.

[8] Orton, A., "Chords, secants, tangents & elementary calculus," *Mathematics Teaching*, 78(1977), 48-49.

[9] Schwarzenberger, R. L. E. and D. O. Tall, "Conflicts in the learning of real numbers and limits," *Mathematics Teaching*, 82(1978), 44-49.

[10] Tall, D. O., "The blancmange function, continuous everywhere but differentiable nowhere," *Mathematical Gazette*, 66(1982), 11-22.

[11] Tall, D. O., *Graphic Calculus* (for BBC compatible computers), Glentop Press, London, 1986.

[12] Tall, D. O., *Readings in Mathematical Education: Understanding the calculus*, Association of Teachers of Mathematics (collected articles from *Mathematics Teaching*, 1985-7).

[13] Tall, D. O., "Graphical Packages for Mathematics Teaching and Learning," *Informatics and the Teaching of Mathematics*, Johnson, D. C. and F. Lovis (eds), North Holland,1987.

[14] Tall, D. O., *Real Functions & Graphs: SMP 16-19* (for BBC compatible computers), Rivendell Software, prior to publication by Cambridge University Press, 1989.

[15] Tall, D. O., P. Blokland and D. Kok, *A Graphic Approach to the Calculus* (for IBM compatible computers), Sunburst, NY, 1990.

[16] Tall, D. O. and S. Vinner, "Concept image and concept definition in mathematics, with special reference to limits and continuity," *Educational Studies in Mathematics*, 12(1981), 151-169.

[17] Sierpinska, A., "Humanities students and Epistemological Obstacles Related to Limits," *Educational Studies in Mathematics*, 18(1987), 371-87.

Visualization and Calculus Reform

Deborah Hughes Hallett

At the beginning college level, visualization is a big part of understanding. Consequently, students who are operating with few mental pictures are not really learning mathematics. Their calculus consists of a vast series of algorithms and a complicated cataloging system which tells them which procedure to use when. The effort put into this kind of teaching and learning is largely wasted: memorized algorithms are soon forgotten and, worse still, such courses perpetuate the idea that math involves doing calculations rather than thinking.

There are many students who can calculate derivatives of extremely messy functions but who cannot look at a graph and tell you where the derivative is positive and where it is negative. Even fewer can tell graphically where a derivative is increasing or where it is decreasing. An alarming number go through a year or more of calculus not knowing whether the graph of $y = e^{-x}$ is above or below the x-axis. Sometimes they remember it is $y = e^{x}$ 'flipped over' — but flipped over what? The x-axis? the y-axis? the line $y = x$? And unfortunately, many believe the only way to know is to memorize the graph.

Similarly, students who can use Newton's method often can't draw a picture to explain the rationale behind it; those finding volumes by integration often can't visualize the discs or shells in the chapter heading.

Of course, we are partly to blame for this state of affairs. Although in a traditional calculus course we may have introduced the derivative graphically and given graphical explanations, we seldom tested graphical ideas. How often have you seen a calculus I final where all the differentiation problems were graphical? How often have you seen a calculus I final where all the differentiation problems were analytical? Part of the problem was that students found graphs hard to draw, and so all too often we made a tacit agreement that they'd only have to draw them in the graphing section of the course. The result is students who react to a more graphical approach by saying "When are you going to stop doing geometry and do math?" Unfortunately, it seems that we have led them to believe that 'real mathematics' consists entirely of the skillful manipulation of x's.

Under a National Science Foundation grant, faculty at Harvard and seven other institutions are designing a new calculus course emphasizing a graphical approach. One of the guiding principles is the 'Rule of Three,' which says that wherever possible topics should be taught graphically and numerically, as well as analytically. The aim is to produce a course where the three points of view are balanced, and where students see each major idea from several angles.

Teaching calculus using the Rule of Three starts with the way functions are represented. If students are to think graphically and numerically, they must be as familiar with functions represented by graphs and tables as by formulas. Since most freshmen have not had much previous graphical work, we start our calculus course with problems asking students to identify functions given graphically. Interestingly enough, even those who have had a BC calculus

Deborah Hughes Hallett is a Senior Preceptor in Mathematics at Harvard University and is co-Director of the Calculus Reform Project based at Harvard and involving the University of Arizona, Chelmsford High School, Colgate, Haverford, Stanford, the University of Southern Mississippi, and Suffolk Community College.

course in high school find this quite hard. Students who want to check their guesses or to investigate the effect of a parameter on a graph use Jim Burgmeier's and Larry Kost's EPIC software. [1]

Before starting a graphical treatment of derivatives, it is important to make sure students know the relative magnitudes of the slopes of the lines in Figure 1. Being able to 'see' the approximate value of a slope is absolutely necessary to understanding derivatives graphically. Again, you will find students who can find the slope of a line given the equation or two points but who need help visualizing the meaning of their calculations.

In our course, graphical differentiation is on an equal footing with analytical differentiation. In Utopia, students should be able to move easily between the graph of a function and the graph of its derivative, between formula and graph, between numerical values and graph. In the real world, these topics are not necessarily easy for students, but discussing them is immensely worthwhile because it builds conceptual understanding rather than just manipulative skill. For example, we believe students should be familiar enough with the graphical meaning of differentiation to be able to decide which of the graphs in Figure 2 is the derivative of the other, without being specially trained to do so.

In addition, we believe it is important for students to be able to write convincing verbal explanations. Problems which get students to reflect on the interpretation of a graph are particularly helpful. The graphs in Figure 3 are an example. They represent the rate two factories are polluting a river, after being ordered to remedy the situation. For each graph, students are asked to write a short paragraph outlining the argument the EPA might make in a case against the factory, and the factory's rebuttal.

A qualitative understanding of whether a quantity is increasing, and if so, at an increasing or a decreasing rate, is frequently far more important than being able to calculate a second derivative. Being able to visualize the difference between increasing at an increasing rate and increasing at a decreasing rate is a crucial component of such an understanding.

Graphs and technology have enabled us to give a much more satisfactory introduction of e than was possible before. The traditional introduction of e^x as the inverse of ln x, which is itself defined as an integral, is usually lost on students who don't like logs and are not yet comfortable with integrals. As a result, the fact that we'd most like them to know — that the derivative of e^x is itself — gets completely buried by the feeling that exponentials are impossibly mysterious. Just why anyone should call a logarithm 'natural' is beyond them. However, with technology it is easy to show students graphically just why e is so important. If the graphs of a^x and $d(a^x)/dx$ are displayed together, students can experiment, and with a little prodding, they 'discover' e as the value of a making the graphs coincide. The data and graphs in Figure 4 were produced by Microsoft Excel™.

■ **Figure 1**

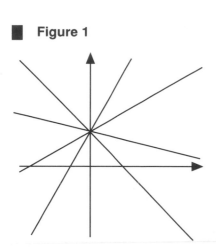

■ **Figure 2**

(Same Horizontal Scales)

■ **Figure 3**

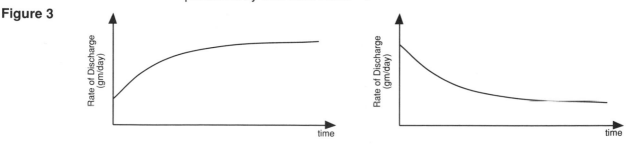

Figure 4a: *An Excel spreadsheet that calculates an exponential function f(x) = a ˣ and its difference quotients, using the approximation*

$$f'(x) \approx \frac{f(x+.001)-f(x)}{.001}$$

The two constants

Base	x-increment	x-value	f(x)=base^x	Difference Quotient	Quotient/f(x)
2	0.1	-2.0	0.25	0.17	0.69
		-1.9	0.27	0.19	0.69
		-1.8	0.29	0.20	0.69
		-1.7	0.31	0.21	0.69
		-1.6	0.33	0.23	0.69
		-1.5	0.35	0.25	0.69
		-1.4	0.38	0.26	0.69
		-1.3	0.41	0.28	0.69
		-1.2	0.44	0.30	0.69
		-1.1	0.47	0.32	0.69
		-1.0	0.50	0.35	0.69
		-0.9	0.54	0.37	0.69
		-0.8	0.57	0.40	0.69
		-0.7	0.62	0.43	0.69
		-0.6	0.66	0.46	0.69
		-0.5	0.71	0.49	0.69
		-0.4	0.76	0.53	0.69
		-0.3	0.81	0.56	0.69
		-0.2	0.87	0.60	0.69
		-0.1	0.93	0.65	0.69
		0.0	1.00	0.69	0.69
		0.1	1.07	0.74	0.69
		0.2	1.15	0.80	0.69
		0.3	1.23	0.85	0.69
		0.4	1.32	0.91	0.69
		0.5	1.41	0.98	0.69
		0.6	1.52	1.05	0.69
		0.7	1.62	1.13	0.69
		0.8	1.74	1.21	0.00
		0.9	1.87	1.29	0.69
		1.0	2.00	1.39	0.69
		1.1	2.14	1.49	0.69
		1.2	2.30	1.59	0.69
		1.3	2.46	1.71	0.69
		1.4	2.64	1.83	0.69
		1.5	2.83	1.96	0.69
		1.6	3.03	2.10	0.69
		1.7	3.25	2.25	0.69
		1.8	3.48	2.41	0.69
		1.9	3.73	2.59	0.69
		2.0	4.00	2.77	0.69
		2.1	4.29	2.97	0.69
		2.2	4.59	3.19	0.69
		2.3	4.92	3.41	0.69
		2.4	5.28	3.66	0.69
		2.5	5.66	3.92	0.69
		2.6	6.06	4.20	0.69
		2.7	6.50	4.51	0.69
		2.8	6.96	4.83	0.69
		2.9	7.46	5.18	0.69
		3.0	8.00	5.55	0.69

Figure 4b: *a chart from the Excel spreadsheet*

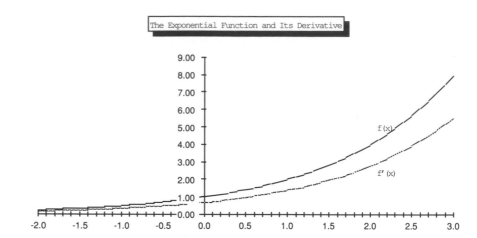

The Exponential Function and Its Derivative

Figure 5: *Direction field of dy/dx = 1/x*

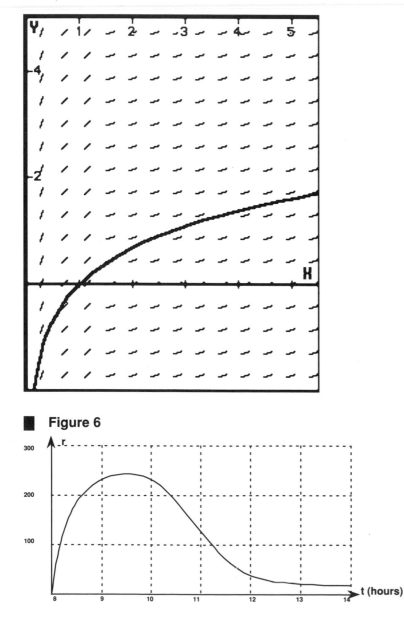

Figure 6

The graphical approach can profitably be continued into integration. Bill McCallum, at the University of Arizona, has suggested using direction fields as a way of introducing the integral which shows clearly why integration reverses differentiation. To 'see' an antiderivative of $f(x)$, look at the direction field of $dy/dx = f(x)$, and trace out any solution curve. For example, $y = \ln x$ is the solution curve shown in Figure 5 [2]. This approach is a wonderful vehicle for explaining the relationship between antidifferentiation and Riemann sums (or Euler's method) because it makes the fundamental theorem almost obvious.

Besides seeing how a function is reconstructed from its derivative via a direction field, students should be able to visualize an integral as an area. Some imaginative examples, such as the following ticket line problem, are in Peter Taylor's book *An Introduction to the Analysis of Functions* [3]. A particularly nice feature of his examples is that they involve approximation and estimation, as well as graphical interpretation. Figure 6 shows the rate at which people arrive to buy theater tickets. The ticket window opens at 9 am and serves an average of 200 people per hour. Students are asked to estimate the number of people in line at 10 a.m., the time at which the line is longest and the time the line disappears. Recently, we asked some students to present the solution to this problem orally and found it an excellent vehicle for developing their understanding of the connections between rates, areas, derivatives, and integrals.

Figure 7a: *Exponential Growth*

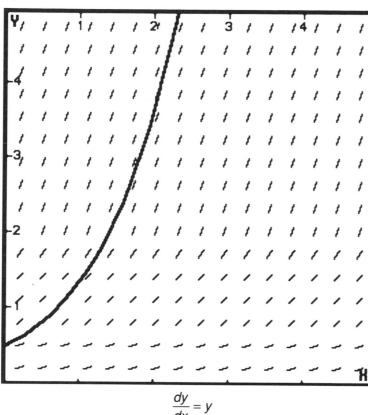

$$\frac{dy}{dx} = y$$

Figure 7b: *Logistic Growth*

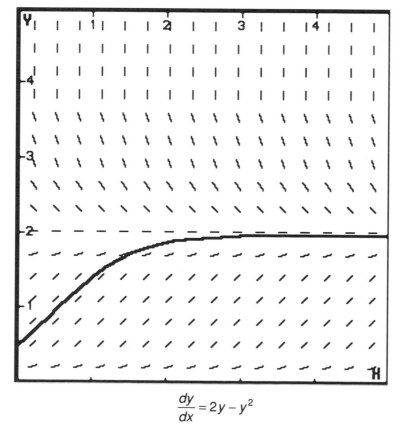

$$\frac{dy}{dx} = 2y - y^2$$

The last topic I'd like you to consider is differential equations. Many students manage to work their way through a chapter, or even a course, on differential equations without ever realizing first-order equations are a set of directions for stepping across the plane. Having students draw solution curves to a direction field should be one of the first things we do in differential equations. Not only does this underscore what a solution curve is, but it also enables students to make qualitative arguments that they would otherwise find impossibly hard. For example, the fundamental difference between exponential and logistic population growth is made much more vividly by comparing the solution curves in Figure 7 [2] than by solving the equations analytically.

The previous examples will, I hope, have convinced you that our students are missing a great deal if they learn to compute but not to visualize. Consequently, I believe we should be redesigning our calculus curriculum to give equal emphasis to the visual, the numerical, and the analytical. However, since the courses are already too full, the only reasonable way to do this is to cut down the number of topics covered. Those remaining should then be taught from as many points of view as possible. Creating a web of connections between different approaches is a big part of understanding. If our students are to emerge from a calculus course with more understanding than they have in the past, they must have more experience with graphs and visualization.

Notice that what I'm advocating is a balance between the graphical, the numerical, and the analytical. I'm not suggesting that our courses should be entirely graphical any more than I'm suggesting that they should be entirely symbolic. A balance is required because it's seeing the links between various approaches that constitutes understanding.

The kind of conceptual understanding that develops from seeing the connection between the algebraic and the graphical doesn't come easily to students, particularly since it has seldom been asked of them in the past. Nevertheless, it's worth striving for. Without making our courses more difficult, we can replace the more torturous analytical procedures with more graphical and numerical work. For example, some of the methods of integration can be replaced by more graphical and numerical problems like the ticket-line problem above. This gives us the opportunity to emphasize the interpretation of results as well as computation, and to start to get students thinking about what they are doing.

Calculus reform will only be a real reform, in my opinion, if it reestablishes thinking as an essential part of learning mathematics. Getting students to think has to be our top priority. Thinking means being able to look at problems from several points of view, one of which must be visual. Consequently, if we are to be successful in reviving calculus, we must move graphs from their supporting, extra-credit role to center stage.

References

[1] Burgmeier, Jim and Larry Kost, *Exploration Program in Calculus*, Department of Mathematics and Statistics, University of Vermont, Burlington, Vermont.

[2] Figures such as these can be constructed with the *Slopes* program by David Lovelock, University of Arizona, or with *Differential Equations* by Herman Gollwitzer, Drexel University.

[3] Taylor, Peter D., *An Introduction to the Analysis of Functions* (unpublished), Department of Mathematics and Statistics, Queen's University, Kingston, Ontario.

Visual Thinking in Calculus

Walter Zimmermann

Of all undergraduate mathematics courses, none offers more interesting and varied opportunities for visualization than calculus. Most of the concepts and many problems of calculus can be represented graphically. Recognizing the importance of graphics in calculus, texts are adorned by numerous figures and diagrams. In many cases, however, these are little more than decorations. In selected cases, diagrams may be used directly as a tool in problem solving, but considering the calculus course as a whole, geometrical reasoning is used inconsistently at best, and the role of visual thinking is not seriously addressed. Few of the examples or problems are designed to develop the student's ability to represent or solve problems graphically. In calculus, we face the paradox that the importance of graphics is widely recognized, but the potential inherent in visualization has not been achieved.

The growth of computer technology, and especially interactive computer graphics, introduces a new and important element into any discussion of visualization. Software is now becoming available which realistically opens up new realms of possibilities for visualization in calculus and other fields. However, it is naïve to believe that technology alone will provide the key to the effective use of visualization. In the absence of a set of guiding principles to relate visualization to the content and learning objectives of the course, any use of computer based or non computer-based graphics is likely to be ineffective.

It seems appropriate to use the term *visual thinking* to describe the aspects of mathematical thought which are based on, or can be expressed in terms of, visual images. It is important to ask how visual thinking skills are learned, how they can most effectively be taught, and how they can be evaluated. However, these empirical and pedagogical questions are not meaningful in the absence of an operational definition of visual thinking. What do we mean by visual thinking? An understanding of this question is not only necessary for the effective use of visualization, but may be a key to a successful program of calculus reform. A concrete way of addressing this question is to ask, *what should a student who is proficient in visual thinking (in calculus) be able to do?* Analysis suggests that the relevant knowledge and skills can be broken down into five categories, which could be called *basic objectives; functional objectives; general objectives; objectives related specifically to calculus; and higher-order objectives.* The classification scheme presented here is essentially pragmatic rather than theoretical. Clearly, the objectives described are interdependent, and other objectives could be mentioned. However, this outline may provide a frame of reference for discussing visual thinking. Let us examine the learning objectives which ought to be part of visual thinking.

Basic Objectives

Basic objectives, which are prerequisites for visual thinking in calculus, include the ability to understand algebra and geometry as alternative languages for the expression of mathematical ideas; to understand the rules and conventions associated with mathematical graphics; to extract specific

Walter Zimmermann is Professor and Chair of the Department of Mathematics at the University of the Pacific. He is an author of educational software for mathematics, a member of the Committee on Computers in Mathematics Education of the MAA, and coeditor of this volume.

information from diagrams; to represent and interpret data graphically; and to plot functions intuitively and with the aid of a computer.

To understand algebra and geometry as alternative languages for the expression of mathematical ideas

One basic aspect of visual thinking in mathematics is the ability to move back and forth between the graphical (geometric) and analytic (algebraic) representation of a problem. Analytic geometry provides the simplest and clearest model of this process. Through the study of analytic geometry, one comes to view the diagram and the equation as two representations of the same object; one comes to see algebra and geometry as two languages for the expression of mathematical ideas. This insight is a transformation of thought which is a prerequisite for visual thinking in calculus. The principles at issue can also be presented by studying elementary functions from an analytical and graphical point of view. In fact, for purposes of calculus, these examples are even more important than the traditional problems of analytic geometry.

To understand the rules and conventions associated with mathematical graphics

Part of learning mathematics is understanding mathematical notation, including formal aspects of the notation and conventions relating to usage and style. Function graphs, geometric figures in two or three dimensions, contour plots, phase-plane diagrams, direction fields, fractal images, and all other kinds of figures make up the language of visual mathematics. Each genre of graphics is designed to convey certain kinds of information and with each is associated certain conventions which must be understood if the figures are to be meaningful.

In *Visual Thinking*, Arnheim [1] writes, "How much do we know about what exactly children and other learners see when they look at a textbook illustration, a film, a television program? The answer is crucial because if the student does not see what he is assumed to see, the very basis of learning is lacking. Have we a right to take for granted that a picture shows what it represents, regardless of what it is like and who is looking? ... Frequently, visual patterns offer difficulties of comprehension ... Careful investigation of what the persons see for whom these images are made is indispensable."

Arnheim is concerned with visualization in a broad context, but his message is pertinent to mathematical visualization. The effective use of visualization in mathematics requires that attention be devoted to an explicit discussion of how to interpret figures and diagrams. In the context of calculus, the most important diagrams are obviously functions graphs and related figures. *With proper preparation*, students can develop insight and understanding of functions through the study of their graphs, but the intuition and knowledge required does not come automatically. It must be learned.

One tends to internalize rules and assumptions associated with graphic presentations. It is sometimes difficult to articulate the formal structure or informal conventions associated with graphics, as anyone knows who has tried to teach a child how to read a map. In order to teach visual thinking, it may be necessary to engage in "consciousness raising" exercises to become aware of the tacit knowledge required to understand mathematical graphics.

To extract specific information from diagrams

One outcome of education in any field is that one comes to see the world differently. The astronomer sees more than I can see when looking at the sky with a telescope, and the trained mathematician can see more in the Mandelbrot set than the nonmathematician. In the development of visual thinking, we should strive to develop students' ability to see mathematical meaning in diagrams of every kind. It is important for students not only to understand a feature of a diagram when their attention is called to it, but to recognize on their own that a diagram may contain information needed for the solution of a problem and to develop the habit of looking to diagrams as a source of such information.

Often the information in a diagram is qualitative in nature rather than quantitative. One may observe from the graph of a function that the function is odd, continuous, monotonically increasing, that it has one inflection point and no critical points, and that it is bounded above and below. Information may be contained implicitly rather than explicitly in a figure. Recognizing information contained implicitly in a diagram depends on appropriate geometrical intuition and knowledge. Students sometimes come to calculus with the notion that information relevant in mathematics is always explicit and quantitative in nature. It is difficult to realize the validity of other kinds of information and to learn to extract such information from diagrams.

To represent and interpret data graphically

Functions as they arise in the sciences are abstractions from data or generalizations of data. Some calculus reform projects stress empirical models and the role of calculus in understanding such models. To understand calculus it is useful to be familiar with the graphical representation of data. The student should be able to represent and to interpret data graphically. The importance of understanding data derives from calculus itself as well as from the empirical sciences. Calculus generates its own data. Numerical integration, the convergence of sequences and series, and generally the study of limits in any context leads to problems which have something of the character of data analysis.

To plot functions intuitively and with the aid of a computer

In the computer age, the relevance of traditional function-sketching skills is an open question. The issue here is the appropriate use of technology. There is the risk that technology will become a crutch and have a negative impact on the development of visual thinking skills. We would not like to see students run to the computer lab or pick up a graphing calculator to visualize $f(x) = 1/x$ or $f(x) = x^2$. What about $f(x) = x/(1 + x^2)$? $f(x) = 4 \sin(x/2)$? $f(x) = (1 - x^2)^2$? Elementary function-sketching skills are visual thinking skills, the same skills a student must be able to draw on intuitively in the process of reading a text, understanding a lecture, interpreting graphics on the computer screen, or solving mathematical problems. An important issue is how students can master these skills, and what role technology can play. One strategy is to say that students should be encouraged to work out the geometrical properties of a function before resorting to the computer, whenever it is feasible to do so. The role of computer graphing software would then be to verify calculations or conjectures about functions, to fill in quantitative details, or to plot functions of a higher order of complexity. Another approach is to use computers or graphing calculators to allow students to explore and experiment with the properties of graphs. However, guidance, feedback, and, eventually, a synthesis of important results must be built into the process.

Since the use of computer software packages to produce function graphs is becoming commonplace, students must not only be able to use these packages but must be able to interpret computer-generated graphics with understanding. The computer does not eliminate the need for understanding the properties of functions in order to graph them. To use a function-plotting package, a student must choose an interval which includes the important features of the function — what is sometimes called a "complete graph." While an appropriate interval may be found by trial-and-error, this method is not efficient and hardly promotes learning. Computer-generated images are subject to misinterpretation on a variety of counts. On a large enough scale, all polynomials whose leading term is the same will look the same, while on a small enough scale, all smooth functions look linear. There is the possibility of errors in entering a function, and (unfortunately) the possibility of bugs in the software. A component of visual thinking is to recognize incorrect or misleading graphics, and to make an appropriate interpretation.

Functional Objectives

The ability to understand how fundamental concepts are represented graphically; to draw diagrams and use diagrams in problem solving; and to use diagrams effectively in proofs are three important objectives we shall call *functional objectives*.

To understand how fundamental concepts are represented graphically

The diagrams associated with fundamental concepts and principles are usually presented in the textbook or by the instructor. However, students often don't fully understand these figures or their implications. One issue that should be recognized is that graphics illustrating basic concepts are typically "generic" graphics; they have many arbitrary elements. To describe the definite integral, any reasonable curve may be drawn, any interval may be selected, any reasonable number of rectangles may be used, and the height of each rectangle (in relation to the curve) may be defined in several ways. What counts is not the details of the diagram but the *structure* of the diagram. The ability to recognize and interpret visual structures is an aspect of visual thinking which may be difficult to teach and learn, but this issue should be given attention if diagrams are going to be used with optimal meaning and effectiveness.

To draw and use diagrams as an aid in problem solving

A diagram is valuable because it may efficiently present the pertinent information in a problem and may show explicit and implicit relationships in a problem which are difficult to see in any other form. On the other hand, the diagram may be difficult for students to draw for many of the same reasons. To draw the diagram, it is necessary to fully and clearly understand what the problem says. Other skills required may include function-sketching skills; algebraic skills required to find points of intersection which may not be specified; and a degree of mathematical intuition, judgment, and self-confidence required to integrate various pieces of information and to resolve apparent ambiguities in the problem.

In teaching students to draw and use diagrams, there are attitudinal problems to be overcome. The paper by Eisenberg and Dreyfus (in this volume) is titled "On the Reluctance to Visualize in Mathematics." They discuss the common fact that "students tend to opt, whenever possible, for an analytical framework to process mathematical information rather than a visual one." Using an algorithmic approach, there is a chance (with some intuition and a little luck) that one may actually get the right answer without fully understanding the problem. While the student may see this as an advantage, it is in fact a serious flaw in the educational process. If we want students to understand the problems they are solving and the techniques they are using to solve them, they should be required to present a diagram whenever a diagram is relevant.

A student may be able to draw the diagram but may not be able to use it to solve the problem. One common obstacle is the inability to extract information contained implicitly in the figure. Another difficulty may be an inability to translate what they see in the figure back into a symbolic form. Finding the appropriate limits of integration in a double or triple integral is a common example. Interpreting diagrams in a problem-solving context is a complex process which involves analytic and synthetic skills and mathematical intuition, insight, and judgment.

Consider the problem of evaluating a definite integral. Perhaps the problem in question involves the definite integral of an odd function over an interval centered at the origin, such as

$$\int_{-\pi}^{\pi} \sin(x)\,dx$$

How many students will recognize on geometric grounds that the integral must be zero? This insight is immediate if one draws the graph of sin(x) on the interval $[-\pi, \pi]$. In other words, the insight is based on visualization. Some would regard the geometric approach as simply a clever trick, an *ad hoc* device which avoids the necessity of "really" doing the problem, but a different viewpoint is that the insight to exploit symmetry is as much a part of mathematics as the ability to use analytic integration techniques. The geometric approach depends on a degree of insight, not mere cleverness. It shows why the result *must* be zero. The straightforward analytic approach depends on no insight, and provides none. It only seems to show that the result *happens* to be zero. The geometric approach is not only valid, but may be a better way of doing the problem, in cases where it applies. Evaluating this integral by the fundamental theorem of calculus, many students, making a sign error, are happy to report that the integral evaluates to 2. The standard calculus course fails to systematically present and consistently reinforce an approach to mathematics which is likely to avoid such nonsense. An important component of visual thinking is the ability to recognize when a problem solution is patently false, on geometric grounds. Visualization is a key ingredient in a more meaningful approach. Examples and exercises which depend on visual thinking should be as common in calculus texts as examples which depend on calculation. In problems where either approach is feasible, both approaches should be presented, and their relationship and relative merits discussed.

To use diagrams effectively in proofs

Barwise and Etchemendy, in this volume, discuss the role of graphics in proofs. They criticize the dogma that diagrams have no role in formal proofs. They also recognize that this is a "heretical view." Whether or not diagrams have a proper place in formal proofs, there is no doubt that diagrams play a heuristic role in motivating and understanding proofs. Examples which come readily to mind are the proof of the integral test for convergence of infinite series with positive terms and the proof of the mean value theorem, based on Rolle's theorem, and proofs of the fundamental theorem of calculus.

There is an important difference between the role of diagrams in proofs and the role of diagrams in problem-solving. A problem diagram describes the *special conditions* of the problem. On the other hand, a diagram associated with a proof must describe the *general case* under discussion. This presents an obvious difficulty because any diagram, being a concrete entity, has its own special characteristics. The special characteristics of the diagram, which do not reflect the general case, must not be used in the proof. A proper interpretation of a diagram, in the context of a proof, requires a tacit understanding of which features of the diagram are incidental and which features are essential. This is a subtle aspect of visual thinking which is difficult for students to apply. (Indeed, this is one of the important difficulties inherent in proofs based on diagrams). We owe it to students to address this issue and to discuss how it works itself out in particular examples.

General Objectives

Objectives in visual thinking which have broad applicability in many areas of mathematics, including calculus, we shall call *general objectives*. These objectives include the ability to understand estimation and approximation in a geometric context; to recognize and exploit symmetry, periodicity, similarity and other patterns revealed in graphics; to understand mathematical transformations visually; and to have in mind a repertoire of important visual images. Although these objectives have wide applicability, examples given will be from calculus.

To understand estimation and approximation in a geometric context

Approximations are at the core of calculus. The derivative and definite integral are defined in terms of a systematic sequence of approximations, and in fact any problem based on the concept of a limit will have this character. Linear approximations to functions, convergence of Taylor's series, numerical integration techniques, and Newton's method are examples of other topics which directly involve estimation and approximation and can be presented graphically. We should come back regularly to the diagrams representing these concepts to extract deeper levels of meaning, to understand how these kinds of diagrams lie at the foundations of calculus and capture the spirit of the subject.

To recognize and exploit symmetry, periodicity, similarity, and other patterns revealed in graphics

Steen [2] has written that "Mathematics is the science of patterns." The role of visualization is to represent the (sometimes abstract) patterns of mathematics geometrically so their structure can be perceived. Some spatial patterns, notably periodicity, symmetry, and similarity, have direct mathematical origins or implications. The work of the Dutch artist M. C. Escher reveals some highly complex patterns of group theory, and the patterns in fractal images reveal a new mathematical world. Escher and fractal images aside, there are aspects of patterns which are important in precalculus, calculus, and related topics. The symmetries of the conic sections are obvious examples. The periodicity of the sine and cosine functions, especially in the context of Fourier series, is another example with interesting visual ramifications.

To understand mathematical transformations visually

In algebra, students become accustomed to problems whose answer is a number ($x = 3$) or perhaps a small set of numbers. In precalculus and calculus they first encounter problems of a higher order of abstraction — problems whose answer calls for a function. They must understand transformations which map functions into functions. Differentiation is obviously an important example of such a transformation. A student's inability to understand the meaning of this kind of question can be a source of disorientation in the first semester of calculus. In any case, it is important to have a firm grasp of this kind of transformation, and the issue can be effectively addressed visually. Among the most important transformations to understand are scaling, inversions, and translations. Given the graph of a function $f(x)$, the student should be able to graph $2f(x)$, $f(2x)$, $f(x + 1)$, $f(x) + 1$, $f(-x)$, $-f(x)$, $|f(x)|$, and $f(|x|)$, without confusion. The behavior of functions which depend on one or more parameters can be understood within this same framework.

To have in mind a repertoire of important visual images

The musician has in mind a repertoire of melodies, and the writer, or any literate person, has a repertoire of phrases, idioms, and fragments of poetry or prose which provide color and texture to language. What does the mathematician retain in the mind? If mathematics is the science of patterns, perhaps the mathematician's repertoire is made up of a set of patterns of various levels of abstraction. Undoubtedly, some of these patterns are retained in visual form. The content of calculus provides a repertoire of simple visual images which are as natural and useful to retain as the musician's repertoire of familiar melodies, and which are an effective means of retaining and integrating one's knowledge and understanding. These may be images associated with concepts (the definite integral), theorems (e.g., the mean value theorem) or selected problems. The reader may find it interesting to review his or her own repertoire. It is very useful for students to consciously develop a repertoire of images to draw on as a guide in problem solving or as an aid in understanding concepts and principles.

Objectives Related Specifically to Calculus

Objectives related specifically to calculus include the ability to understand differentiation graphically; to visualize infinitesimal elements in geometric figures; and to visualize surfaces and related figures in three dimensions.

To understand differentiation graphically

A fundamental aspect of visual thinking in calculus is to understand differentiation from a graphical point of view. Students should be able to sketch the graph of the derivative, the second derivative or an anti-derivative of a function (qualitatively), given a graph of a function. This ability incorporates many ideas connected with the derivative and its relation to critical points, inflection points and other characteristics of curves. The anti-derivative can be generated conceptually in two ways: by thinking about a function whose slope at each point is specified (by the given curve) or by thinking about the "area function" associated with the given curve. An integrated understanding of both of these ideas provides a foundation for an intuitive grasp of the fundamental theorem of calculus.

To visualize infinitesimal elements in geometrical figures

We sometimes fail to communicate to students that the computation of areas and volumes and arc lengths and moments of inertia and, in fact, all applications of the integral calculus depend on a single principle, which is to conceive of an object (or process) as being composed of an infinite number of infinitesimal parts. The essential simplicity of this idea (and the unity of integral calculus) is sometimes obscured by the complexity of some of the details of specific applications. In the interest of developing intuition, it is important, whenever possible, to represent the infinitesimal element graphically and to learn to use the diagram to determine how the infinitesimal element depends on the variables and parameters in the problem. This approach not only provides an intuitive understanding of each particular topic, but provides a basis for understanding the principles underlying all applications of integration.

To visualize surfaces and related figures in three dimensions

Multivariable calculus is, in many respects, a course in three-dimensional geometry. Partial derivatives, directional derivatives, the gradient, tangent planes, level curves and procedures for formulating iterated integrals are all motivated geometrically. Each topic generates its own set of visualization skills, but there are certain aptitudes which enter into almost all of these problem areas. First among these is the ability to visualize a surface, if it is not unduly complex, and to visualize the intersection of a surface with a plane — especially a plane parallel or perpendicular to the z-axis. Some students find this easy to do; others find it difficult. Clearly, this is an area where computers promise to play an increasingly important role.

Higher Order Objectives

Higher-order objectives include the ability to appreciate the beauty of mathematics as revealed through graphics; to interpret real-world phenomena or experiences visually; and to explore and discover mathematical ideas visually.

To appreciate the beauty of mathematics as revealed through graphics

Clearly, this is not a skill in the ordinary sense. Still, an appreciation for the beauty of mathematics, especially as this beauty is revealed visually, is a possible and desirable outcome of visual thinking and a legitimate educational objective. What most obviously comes to mind is the spectacular beauty of some fractal images, but we may overlook the more subtle aesthetics in the simpler images of calculus. There are attractive patterns, symmetries, rhythms, and harmonies in curves and surfaces, in families of curves found in contour plots or phase plane diagrams, and in other topics in calculus. Insofar as possible, the aesthetics of visual mathematics should be shared with students and used to full effect to motivate and inspire their interest in the subject.

Of course, the beauty of mathematics has dimensions which cannot be represented visually. The mathematician sees beauty in an elegant proof or an ingenious approach to solving a problem. As many students have no appreciation at all for the beauty of mathematics, one should consider it a positive step if they can at least appreciate the beauty of mathematical graphics. One can hope that in some cases, the student will be drawn by these graphics into a deeper appreciation for the aesthetic or logical or practical dimensions of mathematics.

To interpret real-world phenomena or experiences graphically

Graphic images serve as an important link between mathematical models and the phenomena of the real world. The ability to interpret simple physical processes graphically is an important aspect of visual thinking. The graphics which describe falling bodies, projectiles, simple harmonic motion, and similar processes, as well as the modes of reasoning which link these graphics to the phenomena, should be clearly understood. Once the relevant concepts are described, a student should be able to interpret graphs and other diagrams which describe physical processes, or to sketch graphs of functions associated (qualitatively) with any simple, familiar phenomenon. Considering the origins of calculus, an understanding of change in the real world is an inseparable part of the subject. Computer simulation and animation have the potential to reveal change vividly and directly. In mathematical visualization, as in scientific visualization, the importance of these tools should not be overlooked.

To explore and discover mathematical ideas visually

The spontaneous exploration of ideas is not only a source of enjoyment, it is full of creative possibilities. Of all the possible modes of thought for the exploration of mathematical ideas, the visual mode is perhaps the most intuitive and accessible. One paper in this volume describes how students can explore and discover patterns in fractal geometry, using computers. In calculus, too, it's natural to think of using computers to facilitate mathematical exploration. One of the themes of the calculus reform movement is an emphasis on "active learning" through exploration and discovery. An important agenda item in the calculus reform program is to design software and develop curricular frameworks to support such exploration. There would seem to be great potential in current techniques of interactive computer graphics, which allow the user to select and move a point around on the screen freely, using the mouse, or to view a three-dimensional figure easily from any direction. An important question is: How can exercises be structured so that students can truly explore, following their imagination freely, with a realistic chance of making meaningful discoveries?

The Landscape of Calculus

One of the major difficulties with the calculus curriculum is that it has become a landscape without landmarks. Some students may see it as a confusing jungle of abstractions. Others may see it as a desert, devoid of intellectual nourishment and apparent beauty. Most likely, many are wandering through a forest, examining one tree, then another and another. They study rules of differentiation, maxima and minima, the fundamental theorem of calculus, implicit differentiation, the natural logarithm, Taylor's series, L'Hôpital's rule, polar coordinates, partial derivatives, and all the rest. All the trees seem pretty much the same. None stands out above the others. There is no clear path through. There are no landmarks by which to find one's position and to gauge one's progress. They never reach a clearing where they can see the forest as well as the trees.

For those who find themselves in a jungle of abstractions, visualization may have the potential to make mathematics simpler and more concrete. For those who feel as though they are wandering through a barren desert, visualization may be able to provide shape and form, texture and color, to reveal the beauty of the landscape. For those who are unable to see the forest for the trees, visualization may provide the needed conceptual perspective. A repertoire of visual images, carefully selected and well understood, may provide landmarks of meaning and progress. Such images can provide structure to the subject as a whole and integrate the various components of a particular topic.

Visual Thinking and Calculus Reform

In any "reformed" calculus curriculum (or in the conventional curriculum), there is a role for visualization and visual thinking, and this role should be explicitly recognized. Visualization can be adapted to almost any framework that might be selected for a calculus course, without conflict. One could argue that visualization and visual thinking should be one of the central elements in calculus reform. Conceptually, the role of visual thinking is so fundamental to the understanding of calculus that it is difficult to imagine a successful calculus course which does not emphasize the visual elements of the subject. This is especially true if the course is intended to stress conceptual understanding, which is widely recognized to be lacking in many calculus courses as now taught.

It is widely recognized that symbol manipulation has been overemphasized and that in the process the spirit of calculus has been lost. Furthermore, the availability of computer algebra systems raises interesting questions about the importance of some symbolic techniques, such as some specialized techniques of integration, in the elementary calculus course. Nevertheless, logical reasoning, symbol manipulation and analytical techniques are part of calculus and cannot be ignored. The challenge is to relate the visual, analytical and numerical dimensions of the subject so that they complement one another in the student's understanding. In most interesting problems, there is a dynamic interplay between algebraic and geometric reasoning, and the goal should be to enable the student to appreciate this interplay and to exploit it creatively.

Of course, technology plays a role. With computers and graphing calculators widely available, it is possible to use visualization in ways that would not be practical otherwise. Computers can be used for numerical and symbolic computations as well as for graphical representations. Because of the multidimensional capabilities of computers, it is possible, in principle, to integrate visualization with numerical and symbolic aspects of calculus. But the character of computers, or the availability of software packages with numerical, graphical, and symbolic capabilities, does not guarantee that this integration will occur. The integration must be built into the structure of the course and into the design of particular topics and problems.

Toward a Taxonomy of Educational Objectives

The organization of this essay was inspired, in part, by the well-known reference work, Bloom's *Taxonomy of Educational Objectives* [3]. The authors of that work described the potential benefits of such a taxonomy. One benefit, it was believed, would be to establish a common vocabulary for the discussion of educational objectives. As they described it, "This should facilitate the exchange of information about ... curricular developments and evaluation devices. ... Beyond this, the taxonomy should be a source of constructive help on these problems. Teachers building a curriculum should find here a range of possible educational goals or outcomes. ... Use of the taxonomy can also help one gain a perspective on the emphasis given to certain behaviors by a particular set of educational plans. ... Curriculum builders should find that the taxonomy helps them to specify objectives so that it becomes easier to plan learning experiences and prepare evaluation devices."

Unlike Bloom's taxonomy, this outline covers a limited subject area and is being put forward tentatively, not authoritatively. Its primary purpose is to stimulate thought and discussion about the issues raised. If readers are motivated to reflect on the nature of visual thinking and perhaps to revise this framework in accordance with their own ideas, the primary purpose of this essay will have been achieved. To the extent that some of the points raised strike a chord with any reader, this outline may serve some of the purposes suggested by the quotations above. The most important of these is "to plan learning experiences and prepare evaluation devices." A high priority should be to develop learning experiences (computer-based and noncomputer-based experiences) to develop visual thinking skills, and to devise or collect examples, problems, exercises and assignments which can be used both to teach and to evaluate such skills.

References

[1]   Arnheim, Rudolf, *Visual Thinking,* University of California Press, 1969.
[2]   Steen, Lynn Arthur, "The science of patterns," *Science*, 616(April 29, 1988).
[3]   Bloom, Benjamin S., ed., *Taxonomy of Educational Objectives*, David McKay, 1956.

Acknowledgement

The work leading to this article was supported in part by the National Science Foundation under grant number DMS-8851255.

Geometric Interpretation of Solutions to Differential Equations

J. L. Buchanan, T. J. Mahar, and Howard Lewis Penn

A striking aspect of introductory differential equations courses is the extent to which analytic tractability dictates the topics covered and the examples considered. Systems of differential equations are a good example. While vector systems are a conceptually straightforward extension of the scalar case, their analytic intricacy either precludes any discussion or relegates them to brief mention. Another example is the treatment of mathematical modeling. Concern for ultimate solvability often results in simplistic models of artificial situations.

This article describes some geometric interpretations of solutions to differential equations which can help students understand the behavior of the solutions. The use of graphical software currently available is illustrated on such standard topics as direction fields and phase plane analysis, the harmonic oscillator, motion in a central force field, power series and Fourier series solutions of partial differential equations. While the emphasis throughout is on qualitative behavior, a solid grounding in theory and a substantial knowledge of solution techniques are still necessary. A student cannot use software without knowing how many initial conditions are required and should not use software if unable to detect inappropriate solution behavior caused by a typographical error.

A section at the end of this paper contains a list of a few software packages that may be used to implement the ideas presented here. The articles [B16] and [B3] have more extensive lists of such software.

Geometric Approach To First-Order Equations

Numerical and qualitative methods can be used to determine geometric properties of solutions to nonlinear differential equations. First-order equations provide the simplest setting in which to study such topics as attractors and bifurcation points.

Without the aid of software it is possible to convey to students what a direction field is, but little more. Interestingly complicated direction fields are too tedious to draw by hand in numbers sufficient for both classroom demonstration and later exercises. All of the general differential equations packages *DE PAD* [A1], *Differential Equations and Calculus* [A2], *MacMath* [A4], *MDEP* [A5], *Microcalc* [A6], *Phase Portraits* [A7] and *Phaser* [A8] have facilities for plotting direction fields.

J. L. Buchanan is Professor of Mathematics at the U.S. Naval Academy. He received his Ph.D. from the University of Delaware in 1980.

T. J. Mahar is Associate Professor of Mathematics at the U.S. Naval Academy. He earned a Ph.D. in mathematics at Rensselaer Polytechnic Institute.

Howard Lewis Penn received his Ph.D. from the University of Michigan in 1975. He is Professor of Mathematics at the U.S. Naval Academy and is a coeditor of the software review column of the College Mathematics Journal.

■ Figure 1

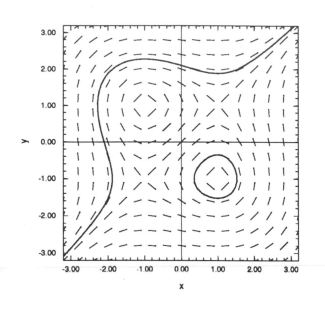

Even a fairly simple equation can lead to an interesting geometry. Let us consider the system of differential equations

$$x'(t) = y^2 - 1$$
$$y'(t) = x^2 - 1 \qquad (1)$$

which has four equilibria, $(-1,-1)$, $(-1,1)$, $(1,-1)$, and $(1,1)$. Upon application of the chain rule, the system (1) becomes a single first-order equation

$$y' = \frac{x^2 - 1}{y^2 - 1}. \qquad (2)$$

Figure 1 shows a direction field for the differential equation (1). With proper classroom display equipment an equation such as this can be investigated in "real time" during class. It may be noted from (2) that the slope is indeterminate at the four equilibria. It is evident from the direction field alone that the behavior of solutions starting near the stable centers $(-1,1)$ and $(1,-1)$ will be qualitatively different from that at the unstable saddle nodes $(-1,-1)$ and $(1,1)$. Thus it is seen that the behavior of a solution to a nonlinear equation is highly dependent upon its initial conditions.

Solution of the differential equation (2) leads to the implicit relation

$$\frac{y^3}{3} - \frac{x^3}{3} - y + x = C \qquad (3)$$

■ Figure 2

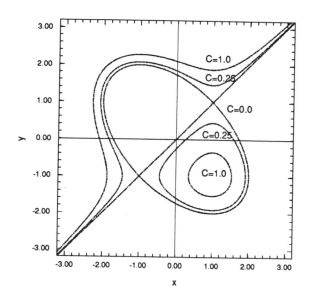

which is not easily made explicit. Some differential equation packages, MDEP [A5] and Microcalc [A6] for instance, can do contour plotting. MDEP was used to produce the solution curves in Figure 1 corresponding to $C = 1$ in (3).

Figure 2 shows the result of decreasing the arbitrary constant C to 0. For $C = 0$ the solution (3) is seen to bifurcate at the two saddle nodes. Clearly this behavior is only compatible with an indeterminate right-hand side and thus such behavior can be exhibited only at equilibrium points of the original system. Having established the curves along which solutions to (1) must travel, one can return to the system and easily determine the direction in which they move by finding the signs of x' and y' in the nine regions formed by the lines $x = \pm 1$ and $y = \pm 1$.

There is no reason that such an example cannot be done in the early weeks of an elementary differential equations course since the only analysis required is the solution of a separable first-order differential equation. Once the idea of stable and unstable equilibria has been grasped, attention can be turned to applications such as

■ Figure 3

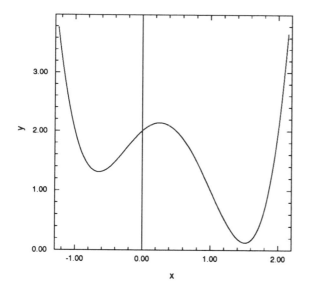

■ Figure 4

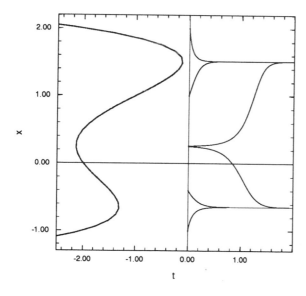

the predator-prey model,

$$x' = a_1 x - b_1 xy$$
$$y' = -a_2 y + b_2 xy \qquad (4)$$

which exhibits a stable center, and the competing species model,

$$x' = a_1 x - b_1 xy$$
$$y' = a_2 y - b_2 xy \qquad (5)$$

in which the equilibrium node is unstable, leading to the demise of one of the species.

Our last example of first-order systems deals with differential equations defined by the negative gradient of a height function. Let $H(x,y)$ denote height above the x,y-plane and consider the system

$$x' = -\frac{\partial H}{\partial x}(x,y)$$
$$y' = -\frac{\partial H}{\partial y}(x,y). \qquad (6)$$

If $(x(t),y(t))$ is a solution curve, the chain rule shows that

$$\frac{d}{dt} H(x(t),y(t)) = -\left(\left(\frac{\partial H}{\partial x}\right)^2 + \left(\frac{\partial H}{\partial y}\right)^2\right). \qquad (7)$$

Thus, the height function is a monotone decreasing function of time away from any critical points; in other words, solutions seek the low ground. Local minima are stable equilibria while local maxima are unstable. This example motivates the term "basin of attraction." Even the one dimensional version of these problems is enlightening. The graph of $H(x) = x^4 - (3/2)(x^3 + x^2) + x + 2$ appears in Figure 3. The minima are attractors and the local maximum repels. Figure 4 shows solution curves $x(t)$ for several different initial conditions and a side view of the height function.

Forced Harmonic Oscillator

Probably the first application of second order linear ordinary differential equations that an instructor will cover is the vibrating spring. The differential equation

$$mx'' + \beta x' + kx = F_0 \cos \omega t \qquad (8)$$

represents the position of a spring with mass m, damping constant β, spring constant k and external force given by $F_0 \cos \omega t$. Since this equation represents an easily visualized system, it is ideally suited for graphical representation. The program *Spring* [A9] animates the solution to this equation. The user inputs values for m, β, k, ω, F_0 and the initial position and velocity. The program determines and displays the analytical solution. It shows animation of the vibrating spring on the left half of the screen while tracing out the graph of the solution in the right half of the screen. The paper [B14] gives a more complete description of this program. Figure 5 shows the screen when the animation is completed for (8) with $m = 1$, $\beta = 2$, $k = 4$, $F_0 = 1$, $\omega = 2$ and the initial conditions $x(0) = 2$, $x'(0) = 0$. This system would be in resonance if the damping constant were 0. *DE PAD* [A1] can also be used to display solutions to vibrating spring problems. This program represents the vibrating spring by a disk which oscillates up and down.

Figure 5

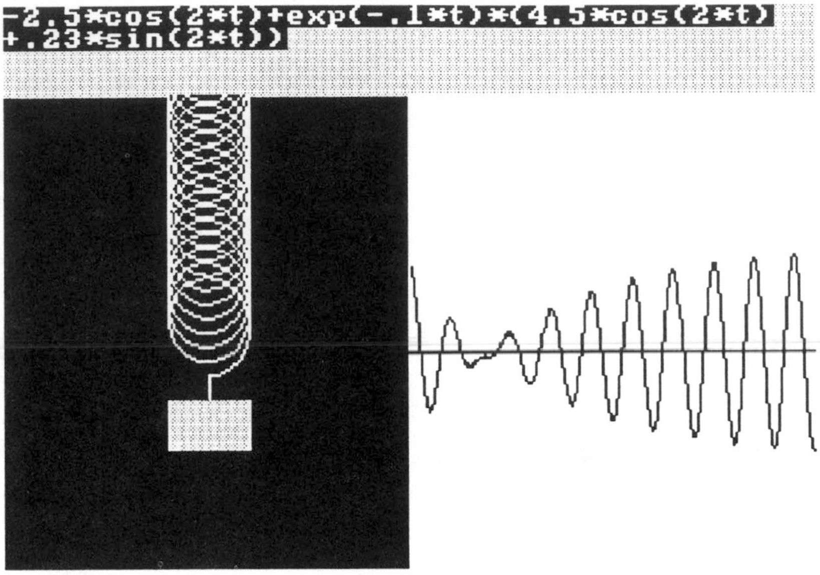

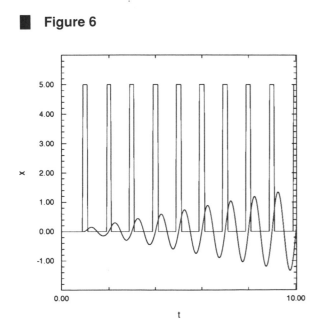

Most texts introduce resonance in undamped mass-spring systems by solving the differential equation (8) with $\beta = 0$. This is the simplest way to get an analytic solution exhibiting oscillations that become unbounded, but students often have little physical intuition about a sinusoidal force applied to an oscillating system. More intuitive is the idea that we simply strike the mass at periodic time intervals. In the resonance case this amounts to striking the mass each time it passes the equilibrium point on the way up.

We write (8) with $\beta = 0$ as

$$x'' + \omega_0^2 x = F(t) \qquad (9)$$

with $\omega_0 = \sqrt{(k/m)}$. The striking force is modeled as

$$F(t) = \sum_{n=1}^{\infty} \frac{u(t-n+\varepsilon) - u(t-n-\varepsilon)}{2\varepsilon} \qquad (10)$$

where u is the unit step function and ε is small relative to the period. This makes solving (9) more difficult analytically, but numerical approximations to solutions are easily plotted. Figure 6 shows the Runge-Kutta approximation with $\omega_0 = 2\pi$ and $\varepsilon = 0.1$. Since the mass is subject to an external force only for brief intervals, it is clear that the critical element for inducing resonance is that the mass has exactly enough time to return to equilibrium before the next strike, thus causing the amplitude of the oscillation to grow indefinitely.

In contrast, Figure 7 shows the result of applying the striking force to a spring with natural angular frequency $\omega_0 = 1.5\pi$. The second and third strikes occur while the mass is heading downward, resulting in a near complete loss of momentum.

The discussion above has been confined to springs which follow Hooke's law and exert a force proportional

Figure 6

■ **Figure 7**

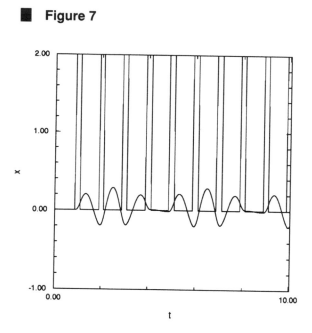

■ **Figure 8**

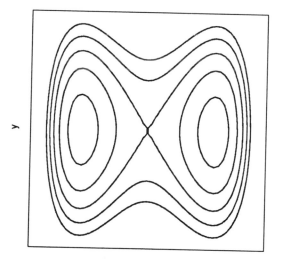

to the stretch. A qualitative analysis of nonlinear springs is easily accomplished by using the conservation of mechanical energy. If $-f(x)$ is the force exerted by the spring when stretched x units beyond its equilibrium length and all other forces are ignored, the equation of motion becomes $mx'' = -f(x)$. If $F(x)$ is an antiderivative of $f(x)$ and x' is denoted by y, the conservation of energy equation becomes

$$\frac{m}{2}y^2 + F(x) = E, \qquad (11)$$

where E is the total mechanical energy. Contour plots of the level curves of the energy function make clear how the solutions behave. If the spring force $-f(x)$ is given by $-(kx + 2ax^3)$, the energy function is $(my^2 + kx^2 + ax^4)/2$. The almost trivial contours are closed curves enclosing the origin and represent periodic solutions. A more interesting energy function results if the force function is $kx - 2ax^3$. While this expression does not correspond to a restoring force, it has interesting features which do occur in real problems. Contours of the energy function $(my^2 - kx^2 + ax^4)/2$ appear in Figure 8. There are two disjoint sets of closed curves around the equilibria at $\pm\sqrt{\dfrac{k}{2a}}$, a separatrix containing the origin, and larger loops encircling all three equilibria. All of the contours except the separatrix correspond to periodic solutions.

Central Force

The vector equations of motion for a particle of unit mass moving in a central force field are

$$r'' = \frac{f(|r|)r}{|r|}. \qquad (12)$$

If the scalar function $f(|r|)$ is always positive, the force is repulsive and the system represents a scattering problem. When the scalar function $f(|r|)$ is always negative, the force is attractive. In the case of attraction, a particle passing close to the force center at the origin will be pulled towards the origin and tend to loop around the force center. As a general rule, these loops will not be closed and the solutions will not be periodic. An orbit for $f(|r|) = \dfrac{-1}{|r|^{3/4}}$ is displayed in Figure 9. The only nonlinear force law which almost always gives closed orbits for small enough energy is the inverse square law given by $f(|r|) = -\dfrac{G}{|r|^2}$. A proof that these solutions are periodic requires a deeper study of analytic and qualitative methods. Orbits for the inverse square law

■ Figure 9

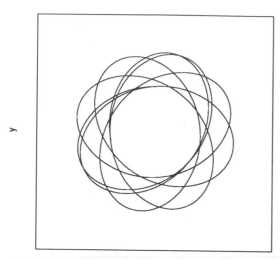

y

x

■ Figure 10

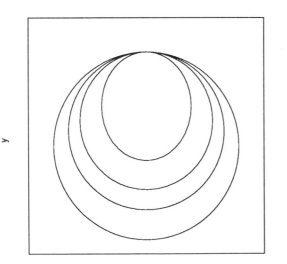

y

x

appear in Figure 10. The launch velocity was varied to generate the different solutions.

Power Series

The differential equation

$$(1+x^2)y'' + y = 0 \qquad (13)$$

has solutions of the form $y(x) = \sum_{k=0}^{\infty} a_k x^k$ where the coefficients of the power series satisfy the recurrence relation

$$a_k = -\frac{(k-2)(k-3)+1}{k(k-1)}a_{k-2}. \qquad (14)$$

MDEP[A5] can plot power series from a recurrence relation such as (14). Figure 11 shows the result of plotting the partial sums $\sum_{k=0}^{n} a_k x^k$ for $n = 2, 4, 6$, and 20. This indicates both the rapidity with which the series converges within its interval of convergence, $-1 < x < 1$, and emphasizes the lack of meaning of the series outside of this interval. The " true" solution, generated by the Runge-Kutta-Felberg numerical method, is also shown.

Partial Differential Equations

When the course covers partial differential equations, Fourier series solutions are usually included. Typically, the book will present the heat equation for a metal bar and the vibrating string as applications. The student goes through a considerable amount of effort to solve a heat equation problem. The result is an infinite series with exponential and trigonometric terms. Most students are frustrated by this solution because they have little hope of picturing it. Computer graphics can be a great help.

A typical heat equation problem might have a metal bar, with insulated sides, of length 100 cm with the left half of the bar at 100 degrees and the right half at 0 degrees. At time $t = 0$, both ends of the bar are placed in ice water. If the diffusivity is taken to be 1, the problem is given by:

$$\frac{\partial U}{\partial t} = \frac{\partial^2 U}{\partial x^2},$$
$$U(0,t) = 0,$$
$$U(100,t) = 0, \qquad (15)$$
$$U(x,0) = \begin{cases} 100, & 0 < x < 50 \\ 0, & 50 < x < 100 \end{cases}$$

■ **Figure 11**

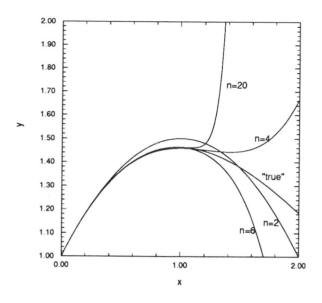

■ **Figure 12**

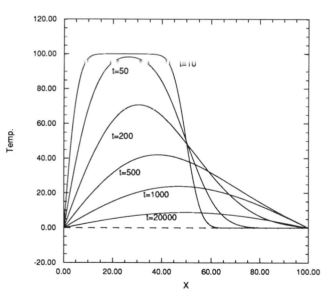

and the solution is

$$U(x,t) = \sum_{n=1}^{\infty} \frac{200}{n\pi}\left(1-\cos\left(\frac{n\pi}{2}\right)\right)e^{-\frac{n^2\pi^2 t}{100}}\sin\left(\frac{n\pi x}{100}\right). \quad (16)$$

Figure 12 contains the graph of the solution to this problem at times $t = 10$, 50, 200, 500, 1000, and 2000 seconds. This graph makes it easy to see the behavior of the solution. One observation that can be made is that the temperature does not change significantly for a while except at the ends and near the middle of the bar. At the exact center of the bar, the temperature remains constant for at least 200 seconds. In the right half of the bar, the temperature increases for at least 500 seconds and then decreases. The hottest point in the bar gradually moves toward the center of the bar.

The paper [B11] contains two examples of the heat equation in which both ends are held at 0 degrees and one example where both ends are insulated. There are also five examples where heat is allowed to radiate at one or both ends. Howard Lewis Penn has produced a videotape showing the solutions to these eight problems [A3].

The vibrating string is the other example of applications of partial differential equations that is normally covered. Let us consider a vibrating string where the length of the string is 2 cm, the string constant $a^2 = 256$ and the initial position is given by

$$Y(x,0) = \frac{1-|x-1|}{48}.$$

The differential equation governing the vibration is

$$\frac{\partial^2 Y}{\partial t^2} = 256\frac{\partial^2 Y}{\partial x^2}. \quad (17)$$

The Fourier series solution is given by

$$\sum_{n=1}^{\infty}\frac{(-1)^{n-1}}{6(2n-1)^2\,\pi^2}\sin\left(\frac{(2n-1)\pi x}{2}\right)\cos(8(2n-1)\pi t). \quad (18)$$

Figure 13 shows the solution for nine values of t ranging from 0 to half the period.

Most students are surprised when they see these graphs. This allows the instructor to remind the students of the assumptions that are made to derive the vibrating string equation. In particular, these assumptions — that there is no external force and that the tension is constant and points in the tangent direction — imply that straight sections of the string have no net force acting on them. These graphs also lead naturally to the D'Alembert solution to the vibrating string equation. The solution,

$$Y(x,t) = \frac{Y(x-at,0)+Y(x+at,0)}{2}, \quad (19)$$

makes the behavior clear.

Figure 13

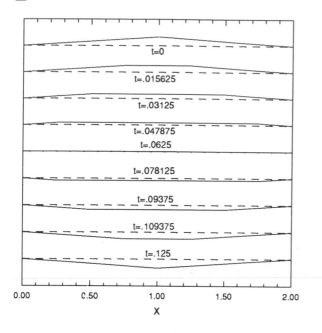

Reference [B13] contains a discussion of several vibrating string problems. The videotape [A3] includes animation of five vibrating string problems. *Spring and String* [A9] contains the solution to the same problems. *Differential Equations and Calculus* [A2] will provide animation to vibrating string problems.

Selected List of Software

The following is a list of some of the software packages and other materials that can be used as aids to the visualization of the ideas of differential equations. The references include a number of reviews of this software. The Compleat Database [B3] lists many other software packages that may be used for differential equations. "A Review of Differential Equations Software" [B16] presents mini-reviews of 20 software packages. Interested readers are invited to contact the authors or publishers listed for price and availability.

[A1]   DE PAD Version 1.0; Steven Rice and Jim Vernor, Queen's University; equipment: IBM PC or compatible with graphics board.

[A2]   Differential Equations and Calculus; Byoung Keum, University of Illinois; publisher: PC-Sig Disk #1072; equipment: IBM PC or compatible with EGA graphics card. Some procedures require a math co-processor.

[A3]   Heat and Vibrating String Videotape; Howard Lewis Penn, U.S. Naval Academy.

[A4]   MacMath version 6.1; John Hubbard and Beverly West, Cornell University; equipment: Macintosh.

[A5]   MDEP (Midshipman Differential Equations Program) Version 2.16; J. L. Buchanan; U.S. Naval Academy; equipment: IBM PC or compatible with 384K memory and a graphics board.

[A6]   Microcalc Version 4.0; Harley Flanders; publisher: MathCalcEduc, 1449 Covington Dr., Ann Arbor, Mi 48103; equipment: IBM PC or compatible.

[A7]   Phase Portraits 2.0; Herman E. Gollwitzer; publisher: contact author at Drexel University for information; equipment: Macintosh.

[A8]   Phaser: An Animator/Simulator for Dynamical Systems; Hüseyin Koçak; publisher: Springer Verlag; equipment: IBM PC or compatible with DOS 2.0 or higher, 256K memory and a graphics board.

[A9]   Spring and String; Howard Lewis Penn; equipment: IBM PC or compatible with graphics board.

References

[B1]   Bridger, Mark, "Review of Differential and Difference Equations Through Computer Experiments," *BYTE*, 11(1986), 63-66.

[B2]   Buchanan, J. L. and Howard Lewis Penn, "Software for Differential Equations," Notes from MAA Minicourse #2, AMS-MAA National Meeting, Atlanta, GA, 1988, available from the authors.

[B3]   Cunningham, R. S. and David A. Smith, "The Compleat Mathematics Software Database," *The College Mathematics Journal*, 19(1988), 268-289.

[B4]   Di Franco, Roland, "A Review of Differential and Difference Equations through Computer Experiments," *The College Mathematics Journal*, 21(May 1990).

[B5]   Greenwell, Raymond N., "A Review of MacMath," *The College Mathematics Journal*, 21(1990), 331-332.

[B6]   Hartz, David, "DEGraph and Phase Portraits," *Notices of the American Mathematical Society*, 36(1989), 559-561.

[B7]   Mahar, T. J., "A Review of DE PAD." *The College Mathematics Journal*, 21(May 1990).

[B8]   Newton, Tyre A., "A Review of MDEP," *The College Mathematics Journal*, 21(May 1990).

[B9]   Newton, Tyre A. and Leonard G. Henscheid, "The Microcomputer as a Teaching Aid for Elementary Differential Equations," Notes for Minicourse given at MAA Sectional Meeting, available from the authors.

[B10] Page, Warren, "A Review of Phase Portraits," *The College Mathematics Journal*, 21(1990), 330-331.

[B11] Penn, Howard Lewis, "Using Computer Graphics to Teach the Heat Equation," *Computers and Education*, 4 (1980), 111-122.

[B12] Penn, Howard Lewis, "Using Computer Graphics to Aid in the Teaching of the Heat Equation with a Radiating End," *Newsletter of the National Consortium of Users of Computers in Mathematical Sciences Education*, 2(1983), 9.

[B13] Penn, Howard Lewis, "Computer Graphics for the Vibrating String," *College Mathematics Journal*, 17(1986), 79-89.

[B14] Penn, Howard Lewis, "Software for the DE Classroom—The Vibrating Spring," *Mathematics and Computer Education*, 21 (1987), 155-158.

[B15] Penn, Howard Lewis, "Differential Equations Software Reviews," *Computers and Mathematics, The Use of Computers in Undergraduate Instruction*, Editors: David A. Smith *et al*, MAA Notes Number 9, The Mathematical Association of America, 1988, Washington, DC.

[B16] Penn, Howard Lewis, "A Review of Differential Equations Software," *Collegiate Microcomputer*, 6(1988), 33-42.

[B17] Riegsecker, John. "Review of Elementary Numerical Techniques for Ordinary Differential Equations", *The College Mathematics Journal*, 17(1986), 182.

Visualization in Differential Equations

Herman Gollwitzer

Visualization for differential equations is most closely aligned with the qualitative aspects of the subject at the present time. This is due in part to the current state of theoretical guidelines and computational tools for visualization associated with differential equations. Rather than attempting to define visualization, we will revert to an account of what has been tried, and what seems to have an element of promise.

We focus mainly on issues related to an introductory course on differential equations. More precisely, we examine the use of exploratory graphics software in such a setting. The presentation is informal and interweaves anecdotal observations, stemming from the author's own experiences at Drexel University, with design issues for the inevitable software that will be used as a delivery system. After describing the setting for such a course, we consider visualization issues for a first-order equation $y' = f(x,y)$ and a two-dimensional system $x' = f(t,x,y), \; y' = g(t,x,y)$. Finally, two examples are presented that illustrate unexpected behavior for associated numerical processes and provide opportunities to discuss difference equations. The paper closes with a brief discussion of the software used to illustrate points in the paper and obstacles to visualization.

The Setting

Drexel University has had an active microcomputer program for all students and faculty since the Macintosh was first introduced. Our visualization experiments in a differential equations course date back to the early 1970's when Drexel received a National Science Foundation COSIP grant. Serious efforts toward using exploratory graphics software in an introductory differential equations course began in 1987.

The traditional introductory course on differential equations has been criticized as being methods-oriented. This is understandable when one compares the mathematics accessible to second or third year students with the complexity of the subject. A student usually approaches a differential equations course with several terms of calculus, little or no geometry or geometrically oriented reasoning abilities, and maybe linear algebra. An introductory course is created by choosing several topics from the following list:

1. First order equations including methods and applications;
2. Second order linear equations and applications from mechanics or circuits;
3. Power series methods for second order equations;
4. Higher order linear equations;
5. Two-dimensional systems including stability concepts and applications;
6. Matrix systems;
7. Numerical methods;
8. Partial differential equations with separation of variables;
9. Laplace transform techniques.

Herman Gollwitzer studied under Professor Yasutaka Sibuya at the University of Minnesota and is currently at Drexel University. In addition to serving as the Assistant Department Head for Graduate Programs, he is involved in developing education software for mathematics. One application, Phase Portraits, *received a "Distinguished Software" award in the 1988 EDUCOM/ NCRIPTAL competition.*

The design of an introductory course and its subsequent presentation has not changed much over the last thirty years. Current practice reflects the analytic heritage of the subject and the lack of comprehensive graphical tools. The topics in this list have not been examined in detail to see where visualization techniques might be fruitful, and so efforts are fragmented. Recent visualization tools provide reasonable support for solution plots and the presentation of the direction field of a scalar equation or a two-dimensional autonomous system. They offer graphical complements to topics 1, 2, 5, and 7. We intend to discuss tools of this nature even though they address, in part, a need that goes back a hundred years or more.

The Scalar Case

We use the equation $y' = -y+x$ for most of the discussion and also mention that parametric studies are useful when explaining classes of equations, or presenting a graphical proof of the theoretical fact that solutions depend continuously on parameters.

■ **Figure 1**

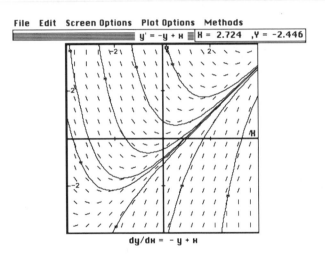

File Edit Screen Options Plot Options Methods

$y' = -y + x$ H = 2.724 ,Y = -2.446

dy/dx = - y + x

The direction field is an important tool for motivating students to think about the visual component in differential equations. Various solutions can be approximated, displayed, and discussed after initial conditions are specified using a keyboard or mouse. The display in Figure 1 is reduced in size for editorial purposes.

These graphic records address issues that many people consider obvious for software of this nature. Screen space is a scarce commodity, and it is tempting to put up too much for viewing. It is difficult to predict when screen features are useful or a hinderance. For instance, the direction field, axes, and rulers may or may not be important for interpreting a differential equation. Accordingly, it should be possible for them to be removed or shown at any point without affecting existing solution plots. The freedom to change almost every presentation feature, at any time, without interfering with existing solution plots, is an important feature in software that is intended to be an exploratory graphics tool for differential equations.

The convenience of being able to look at the direction field or solution plots is not the only dividend of having an exploratory graphics tool available, but more research must be carried out to find meaningful settings where visualization enhances the curriculum. Here is a simple example that illustrates how visualization complements the analytic component, and increases one's awareness of how parameters affect the behavior of solutions. The solution of $y' + ky = x$, $y(0) = b$ is

$$y(x,b,k) = (b + k^{-2}) \, exp(-kx) + (kx - 1)k^{-2}$$

■ **Figure 2a**: $y' = -5y + x$

when k is positive. Students who leave the problem at this point have only demonstrated their understanding of calculus and gain no appreciation of how the parameter k affects the solution. Even though the explicit dependence on k is somewhat complicated, a few snapshots of the associated direction field in Figures 2a, 2b, and 2c for $k = 5$, 1, and 0.1, respectively, stimulates class discussions and leads to conjectures concerning the solutions when k is large

■ Figure 2b: $y' = -y + x$

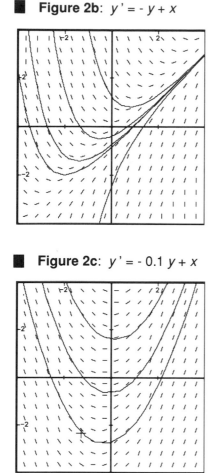

■ Figure 2c: $y' = -0.1\,y + x$

■ Figure 3

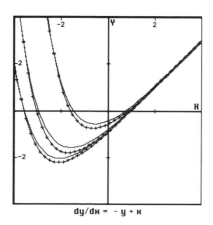

$dy/dx = -y + x$

or small. After that, students can try to use the analytic expression to get an understanding of the solution for large k and x.

The situation changes dramatically when k is small and x is moderate. The near parabolic behavior so obvious from the direction field and plots is not apparent when one examines the analytic expression. Even more striking is the apparently smooth transition to a parabolic solution as $k \to 0$. Such a transition is not apparent from the solution expression. The exponential in the analytic solution can be expanded in a power series to convince a student of this point, but the graphical image has already convinced him and a power series expansion isn't always the most persuasive argument at this level. At this point, an instructor has the clear option of going through the asymptotic details or starting another topic.

It takes time to prepare and present lectures that include visualization. The author's own experience suggests that an entirely different course starts to emerge when visualization is used routinely. More time is spent discussing concepts and interpreting graphics, and less time is devoted to solving numerous routine problems. Students not accustomed to thinking about pictures don't always appreciate the points being made as quickly as we might prefer, and an instructor must be careful to offer a balanced approach to the subject. Current textbooks are not written with visualization in mind, and an instructor must make some hard decisions. He must produce supplementary notes and problems for a class, and make decisions as to what will be omitted from an already crowded syllabus. Not all instructors feel comfortable in doing this and investing the time necessary to make it work. This situation should improve when appropriate lectures and problems are devised and included in textbooks.

The nature of solution plots is also an important issue. Discrete variable methods are used to calculate approximations to initial value problems. It isn't too important to emphasize algorithms when numerical methods are not included, provided that good adaptive algorithms are used. The adaptive versions reduce the effects of discretization and free the user to examine the differential equation. For conceptual purposes, it is important for all students to understand that algorithms generate a sequence of points (x_k, y_k) that approximate the true values $(x_k, y(x_k))$. A default presentation mode is to use linear interpolation between successive pairs of (x_k, y_k). Beginning students who don't appreciate this point get confused as to how the computer "knew the formula" for the solution so that it could be sketched. This can be particularly frustrating for students who may have just learned that most differential equations don't have closed form solutions. Visualization can help students appreciate this point too if different plotting modes (e.g., points, lines, or points and lines, varying pen widths and colors, etc.) are included.

The numerical methods used to generate the approximate solutions present new challenges. There are several standard methods discussed in an introductory course: Euler, Modified Euler, Heun, Adams, and Runge-Kutta. Almost all textbook authors restrict their attention to the case when the stepsize is fixed. Some of these methods are known to be inadequate when compared with those that change the step size according to a strategy that attempts to control errors. It is important that many different fixed stepsize methods be made available so that students can experiment for themselves and see some of the differences. Current visualization tools do not work particularly well in this area because computer screens are not precise enough to display floating point phenomena that ranges over several orders of magnitude. A small step size may lead to graphs that don't reveal truncation or roundoff errors. One way to demonstrate errors is to use the same initial value problem and compare a given method with an accurate method that acts as a reference curve. Figure 3 records the results for the equation $y' = -y + x$ when the Euler method, with stepsize $h = 0.2$ and plot mode "points and lines," is compared with an adaptive fourth order Runge-Kutta method using "lines" as a plot mode.

■ **Figure 4**

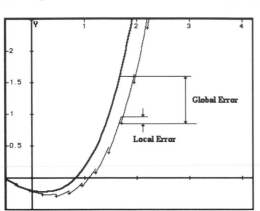

A different type of comparison can be made after the more detailed concept of local error has been introduced. The local error is the difference $u_k(x_{k+1}) - y_{k+1}$ between a solution u_k of the initial value problem $y' = f(x,y)$, and the value y_{k+1} computed from the algorithm under consideration. This is to be contrasted with the global error $y(x_{k+1}) - y_{k+1}$. It is known that a small local error does not necessarily imply a corresponding small global error. With the proper choice of equation, method, and stepsize, one can show the local error and compare it with a reference curve generated with an adaptive fourth order Runge-Kutta method. One approach is to plot the solution u_k over $[x_k, x_{k+1}]$ and show a mark associated with the Euler method. The reference curve is plotted later, and Figure 4 shows the result.

This example helps to make the point that the limited resolution of a screen and the nature of the phenomena may make visualization difficult. Care must be taken to ensure that meaningful examples are chosen, for the author remembers several occasions when ill-considered examples led to defective presentations.

Two-Dimensional Systems

The system of differential equations $x' = f(t,x,y)$, $y' = g(t,x,y)$ has much richer possibilities in terms of visualization. A meaningful discussion of this system requires previous knowledge of two-dimensional matrix systems, eigenvectors, and eigenvalues, because linearization about a critical point is a major analytical tool. Three-dimensional plots or various orthographic projections come to mind immediately, but scaling problems and the general unavailability of color make three-dimensional displays difficult at present.

■ **Figure 5**

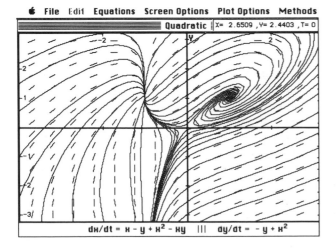

Although it may seem that this material is more suitable for an advanced course, we have found that students can understand many of the points and are reasonably receptive to the visualization component. As an aside, the author's Drexel colleague Chris Rorres has used visualization successfully in graduate courses on differential equations for many years. More recently, he used the software presented here to develop an extensive series of striking phase plane plots that have been made part of a HyperCard® stack. Included are examples of bifurcation and singular perturbation phenomena. The phase plane plot of y *vs.* x provides a rich environment for talking about autonomous systems, but plots of y *vs.* t or x *vs.* t can also be of value. The former is illustrated in Figure 5 for a quadratic system that has a saddle point, node, and spiral point.

Figure 6

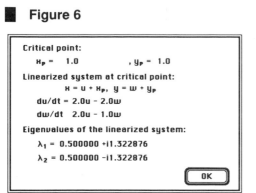

Critical point:
$x_p = 1.0$, $y_p = 1.0$
Linearized system at critical point:
$x = u + x_p$, $y = w + y_p$
$du/dt = 2.0u - 2.0w$
$dw/dt 2.0u - 1.0w$
Eigenvalues of the linearized system:
$\lambda_1 = 0.500000 +i1.322876$
$\lambda_2 = 0.500000 -i1.322876$

OK

The pictures are more striking than in the scalar case, but the level of sophistication needed to understand what is going on has risen too. Both students and instructors have to work harder, but many students enjoy the change of pace this topic offers. It also provides a bridge to more advanced and recent topics such as chaos and fractals.

The most interesting behavior in the phase plane occurs near critical points. It is usually a tedious task to look for critical points and classify them. Using the quadratic system as an example, a search rectangle can be specified with a mouse. Then a search is conducted for a critical point. If found, optional linearization information can be displayed. This is illustrated in Figure 6 for a search rectangle that included part of the first quadrant.

Figures 5 and 6 barely introduce the type of phenomena one can study for two-dimensional systems.

Numerical Concerns

Even a brief discussion of numerical algorithms can lead to phenomena which seem counter-intuitive at first glance, and it is important for instructors to be aware of these pitfalls. A simple initial value problem $y' - 1 - 2xy$, $y(0) = 0$ leads to unusual behavior regardless of the elementary algorithms considered, while the linear oscillator $x'' + \omega^2 x = 0$ and Euler's method combine to give opportunities to discuss linear difference equations and the practical problem of interpreting graphics.

Current computing power enables one to carry out extensive calculations in a short time. Traditional hand calculations rarely revealed interesting phenomena because so few were carried out. Although it would seem that more is better, there is price to be paid — a careful interpretation of the presented approximations. The initial value problem $y' = 1 - 2xy$, $y(0) = 0$ provides a simple example to illustrate this point. An expression for the solution is

$$y(x) = e^{-x^2} \int_0^x e^{t^2} dt,$$

Figure 7

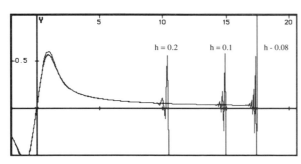

$h = 0.2$ $h = 0.1$ $h - 0.08$

so that $y \to 0$ as $x \to \infty$. The solution restricted to the interval [0,20] should capture most of the interesting behavior, and that is our main interest here. Introductory courses emphasize Euler's method; two second-order Runge-Kutta methods that go by the names of Heun and modified Euler; and the famous classical fourth-order Runge-Kutta method. There are others. Rather striking behavior occurs when Euler's method is used with step sizes of $h = 0.2$, 0.1, and 0.08, and the results are given in Figure 7.

Figure 8

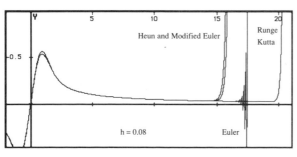

Heun and Modified Euler

Runge Kutta

$h = 0.08$ Euler

The different approximations seem close beyond the initial hump and before the onset of the violent oscillations, and decreasing the step size just seems to postpone the inevitable. The static nature of this picture can't even begin to capture the astonishment one experiences when this example is first seen in real time on a system with good graphics. It provides an excellent example for motivating many aspects of the numerical solutions of differential equations. It is tempting to use higher order methods, but they don't fare much better and fail in an even more unexpected way. These are presented in Figure 8. The fourth-order method takes a little longer to fail, and no reasonable behavior on the interval [0,20] seems possible without reducing the step size even more.

Although contrived, this exercise brings up several useful points. Simple numerical schemes may exhibit behavior that has nothing to do with the equation in question. A graph or table of values may be misleading — the last figure is a good example. An unsuspecting person may put too much trust in the famous Runge-Kutta method and assume that the solution does blow up. Only external qualitative information can provide some reliable indicators when simple methods are used for purposes of illustration. Other questions arise. How does one know when an approximation is acceptable? Why did the Euler approximations oscillate and the others merely increase without bound? Part of the answer lies in a more careful analysis of the factors that cause single step schemes to fail.

The linear oscillator $x'' + \omega^2 x = 0$ is a standard example in elementary courses. A time differentiation shows that the "energy" function $E(t) = x'^2 + \omega^2 x^2$ is constant along a solution — a mechanical interpretation is that total energy is conserved. The locus $E(t) = \mu$ describes an ellipse in the xy-phase plane when $y = x'$, but the solutions spiral into the origin whenever the model equation has an additional positive damping term cx'. So much for the analytic aspects.

Figure 9

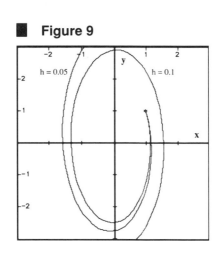

Assume that $\omega = 2$. With step size $h = 0.1$, Euler's algorithm reveals an unexpected behavior — the trajectories are demonstrably not ellipses. All solutions shown in Figure 9 start at $(1,1)$. It takes longer for the spiral behavior to be observable when h is reduced, but it persists. The purely imaginary nature of the eigenvalues of the associated two-dimensional system can lead one to suspect that roundoff errors contaminated the approximation process and caused the resulting spiral effect. Before going on to the underlying difference equations, someone suggests that maybe the problem disappears when positive damping is introduced into the model. Figure 10 shows that outward spirals occur even when the model equation $x'' + cx' + 4x = 0$ has positive damping. For sufficiently large damping c, the trajectories appear to be elliptical with axes that do not line up with the coordinate axes, while the anticipated inward spirals do occur when c is large. The three types of behavior are shown using the same initial point $(1,1)$. The wide curve depicts periodic behavior with positive damping for the Euler method.

Figure 10

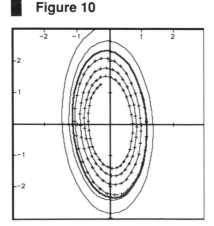

The reasons are revealed only when one studies the associated difference equation. Solutions of the differential equation involve exponentials that contain the roots of the associated quadratic polynomial or, equivalently, the eigenvalues of the associated companion matrix. The underdamped case leads to inward spirals in the phase plane. If $t_n = nh$ and $(x(t_n), x'(t_n)) \approx (x_n, y_n) \equiv u_n$, the Euler algorithm can be expressed as $u_{n+1} = A(h,c,\omega)u_n$ for a matrix A of size 2. A solution u_n of the difference equation is of the form $pr_+^n + qr_-^n$ for certain column vectors p and q. The quantities r_+, r_- are complex and greater than 1 in modulus for small c, and this accounts for the outward spiral behavior. The same calculations show how large c must be to predict the elliptical behavior.

This simple example shows some of the differences between solutions of differential equations and the solutions of associated difference equations. The fact that some positive damping does not change the outward spiraling behavior is counter-intuitive. These considerations lead to many questions for discussion and further analysis. How does the truncation error affect the approximations? Do similar results hold for other methods of approximation, or are there intrinsic difficulties associated with this equation? Problems of this nature can and do come up in the classroom when students and instructors have access to exploratory software for differential equations. This is another reason why it is so important for instructors to become familiar with these aspects of differential equations so that their presentations can be more varied and informative.

Obstacles to Visualization

What prevents visualization from being used more in differential equations? Some reasons were given earlier. Textbooks don't offer enough discussions, examples, and exercises. Software to support these activities is starting to appear, but developers must contend with diverse computing and graphics environments. Students and schools have not developed strategies for obtaining and using software. School administrators are hard-pressed to find the funds necessary to acquire and maintain hardware and software. Time constraints for students and instructors will cause intuitive software tools to be developed, and graphics will only be part of the final product. Assessment remains an open question. It is difficult to devise traditional test questions concerning visualization. Instructors may feel reluctant to use something that can't be tested. Better questions will be devised, but the author feels that the purpose of visualization is insight, and not necessarily assessment.

The Software

The two pieces of software used in the illustrations were developed by the author with invaluable programming and design support from Drexel's Software Development Group. The scalar case is presented in *Differential Equations*, while *Phase Portraits* provides two-dimensional support. The latter application received a "Distinguished Software" award in the 1988 EDUCOM/ NCRIPTAL competition. They are distributed through Drexel University and a national distributor. Software reviews for differential equations are starting to appear [3], and the *Notices of the American Mathematical Society* contain regular articles of interest. Potential developers should read about the efforts of others [1], [2].

References

[1]   Barwise, Jon, and John Etchemendy, "Creating Courseware," *Notices of the American Mathematical Society*, 36(1989), 32-40.
[2]   Chabay, Ruth W., and Bruce A. Sherwood, *A Practical Guide for the Creation of Educational Software*, Center for Design of Educational Computing, CDEC Technical Report No. 89-06, April, 1989, Carnegie Mellon University.
[3]   Hartz, David, "DEGraph and Phase Portraits," *Notices of the American Mathematical Society*, 36(1989), 559-561.

Visualization in Intermediate Analysis Courses in an Integrated Computer Environment

Roland di Franco

In one of the episodes of "The Ascent of Man," Jacob Bronowski sits before a terminal observing a wire-frame image of a human skull which has been reconstructed from a bone fragment. He discusses the characteristics of the skull as he rotates the image on the screen and considers alternative hypotheses about how the original skull may have been shaped. The lecture represents a paradigm for visual thinking. The implications of assumptions can be displayed visually and quickly. Shape and viewpoints of the object can be altered with ease. The focus of attention is on what the images tell us about the subject under investigation.

When Bronowski presented these lectures in 1976, animated wire-frame images seemed to be an impressive achievement. Since that time, computer graphics technology has grown in its capability to create high quality images for both static and dynamic simulation. Today, using state-of-the-art simulation technology, that same skull could be seen as a textured color image of the bone structure with instantaneous viewer control of the image. It is now possible to simulate with much greater realism an actual physical model of such a skull.

Visualization within mathematics involves a range of tasks which require some of the more advanced methods of computer imaging. For the simple tasks, there is need for plotting curves in the plane or in space and line drawings or wire frame images of simple surface graphs. For more complex objectives, there is a need for images with multiple surfaces. To meet these needs, a mathematician requires a computer environment which includes a graphics library with a variety of plotting types and the capability for both static and dynamic displays.

Parallel with the growth of capabilities in computer graphics, there has been rapid development of numerical methods for solving mathematical problems. Libraries with of a wide variety of numerical solvers are now available. In an integrated environment, representing the data in an image provides both a visual assessment of assumptions of the associated model and an evaluation of the accuracy of the underlying numerical routine. This is the same process Bronowski used when he examined alternative hypotheses for the shape of the skull.

Although computer algebra systems were once so large that they would consume the total capacity of a single computer, such systems have now been made so compact that they can coexist with other computer libraries. Such systems are attractive since they promise to take the drudgery out of algebraic computations. But there are hazards in the construction of symbolic algebra programs which create the need for evaluating the correctness of such programs. Again, an environment in which the user can display an image related to the algebraic expression brings visual thinking to bear on the problem.

Roland di Franco is Professor of Mathematics at the University of the Pacific. As a visiting professor at Harvey Mudd College he directed Mathematics Clinics on the mathematics of video simulation and on the industrial use of computer algebra systems. He has also been an industrial consultant on the construction and use of image databases for training simulations.

The effort to create an integrated computer environment which includes libraries of computer graphics routines, numerical solvers and symbolic algebra systems has been under way for some time. Some efforts such as MACSYMA, MAPLE, and SMP began with a core of a computer algebra system and expanded to add numerics and graphics. MATHLIB, a project begun at Harvey Mudd College, combined the TEMPLATE computer graphics system, from the TGS Corporation, with a library of numerical routines and a computer algebra system with a powerful parser and a symbolic differentiator. The "new kid on the block" is, of course, *Mathematica,* which attempts to accomplish the same agenda by combining a new computer algebra system with high quality graphics images and a numerical library.

It is not the intention of this paper to make a careful comparative analysis of these systems. Rather the effort here is to report on how such integrated systems might be used to address issues of visualization in the classroom. The author has been experimenting with the use of one such system, MATHLIB, in courses in differential equations and multivariable calculus. Interactive computer demonstrations have been used in lectures and were followed by computer assignments which integrate graphics into standard textbook problems. This paper discusses two examples in which visualization is a key element. The first example is the modeling of systems of differential equations which depends on an integrated use of graphics and numerical computations. The second example from multivariable calculus is the display of surfaces and their tangent planes which combines graphics with symbolic algebra systems.

Visualization in Modeling with Differential Equations

One of the major objectives in differential equations is teaching the process of mathematical modeling. In modeling, some physical phenomenon is being analyzed in an effort to understand and predict its behavior. Access to numerical solvers has provided a new capability to utilize nonlinear models. When the numerical results are immediately transferred to visual displays, it is easy to compare the predicted linear and nonlinear behaviors. The undamped simple pendulum is discussed as an example of the comparison.

On the other hand, students now need to learn about numerical characteristics of mathematical models. Sensitivity to initial conditions is one of these characteristics. Using the Lorenz equations and a solver with high accuracy, it is easy to visualize this phenomenon.

The simple pendulum is one of the most common examples of mathematical modeling in differential equations. Numerous alternative assumptions can be considered. If the oscillations are small, the equation can be linearized. For the undamped pendulum, the differential equations are

$$mL\Theta'' + mg \sin \Theta = 0 \quad \text{(Nonlinear)}$$
$$mL\Theta'' + mg \Theta = 0 \quad \text{(Linear)}$$

where $\Theta(t)$ is the angle measured from the downward vertical, m is the mass of the pendulum of length L and g is the acceleration of gravity. Textbooks of the subject place great emphasis on the linear equation because it can be solved algebraically. The effects of changes in initial conditions and constants can be shown by comparing algebraic solutions, or more effectively, plotting the graphs on a single axis system. Solving the nonlinear equation requires the use of a differential equations solver. The comparison of the data for the

Figure 1. *Phase diagram for the non-linear undamped pendulum*

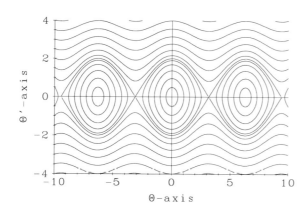

Figure 2. *Phase diagram for the linear undamped pendulum*

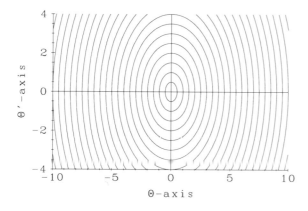

Figure 3. *Comparison of the angle, Θ, for the non-linear (solid line) and linear (dashed line) models of an undamped pendulum for $\Theta = \pi$ and $\Theta' = -0.1$*

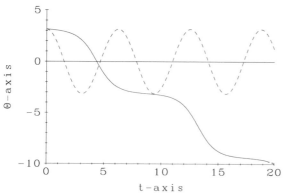

linear and nonlinear solutions is a burdensome task which is relieved by converting the data to visual displays.

To use the solver in MATHLIB, the equations must be translated into a system of first-order differential equations. With $x_1 = \Theta(t)$ and $x_2 = \Theta'(t)$ the nonlinear equation becomes

$$x_1' = x_2$$
$$x_2' = -\omega^2 \sin x_1$$

with $\omega^2 = g/L$.

The linearized version of these equations is:

$$x_1' = x_2$$
$$x_2' = -\omega^2 x_1$$

The resulting phase diagram of the nonlinear and linear systems are found in Figures 1 and 2, respectively.

For small oscillations, both models predict undamped oscillation, but the nonlinear model predicts an unstable equilibrium position when $\Theta = \pi$ and $\Theta' = 0$, namely when the pendulum is straight up. If there is a nonzero velocity at this position, say $\Theta' = -0.1$, the nonlinear model predicts continual circular motion while the linear model predicts oscillatory movement. This can be seen in Figure 3 which shows the evolution of the angle Θ in time, for the two cases.

Besides facilitating the use of nonlinear models, the use of solvers provides an effective way to teach students about a property of mathematical models called sensitivity. A sensitive model is one for which small changes in the initial values result in a new solution which is no longer "close" to the original solution. Older elementary textbooks of differential equations rarely discuss sensitivity except, perhaps, as a theoretical issue. A classroom demonstration which displays numerical solutions to the Lorenz equations for initial conditions which are a "hair's width" apart quickly convinces students of the existence of sensitive systems which have realistic applications.

The Lorenz system models convective air flow between overhead clouds and the land surface and is described by the equations:

$$x' = -\sigma x + \sigma y$$
$$y' = rx - y - xz$$
$$z' = -bz + xy$$

Here x represents the amplitude of the convective air current, y is the difference in temperature between the rising and falling air currents, z is the deviation from normal of the temperature of the system and σ, r, and b are dimensionless physical constants. See [1].

To observe the behavior of sensitivity, a solver with a high degree of accuracy should be selected. The DEQSOLVE facility within MATHLIB is a solver constructed around ODEPACK, a research tool designed at Lawrence Livermore

Figure 4. *Phase plot of the Lorenz equations*

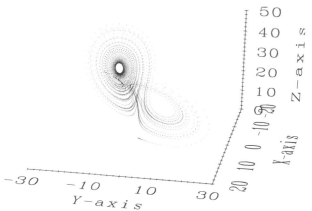

National Laboratories. Because graphics has been integrated into the environment of MATHLIB, the equations can be solved and displayed using a fifteen line command program. Using the parameter values $\sigma = 10$, $r = 28$, and $b = 8/3$, initial conditions of $x = y = z = 0.1$ and solving for 5000 points, we obtain Figure 4, which shows a plot of the x-y-z curve of the solution.

To show the sensitivity to initial conditions, the system was solved first for $x_0 = y_0 = z_0 = 10^{-5}$ and then for $x_0 = y_0 = z_0 = 10^{-3}$. The function $x(t)$ was plotted for these two initial conditions on the same axis. Figure 5 shows the marked difference of the two solutions.

Were the numerics to stand alone, the user would be faced with comparing lists of x-values for the two initial conditions. This would be complicated by the fact that the adaptive grid of t-values generated by the program for each case might differ. The visual image of the data conveys the comparison immediately.

Classroom demonstrations of nonlinear models and the sensitivity of models gives students an understanding of the limits of mathematical knowledge. Discussions with students generate questions which are in fact current issues in mathematical research.

Visualization in Multivariable Calculus

Figure 5. *Comparison of $x(t)$ for $x_0 = 10^{-5}$ (solid line) and $x_0 = 10^{-3}$ (dashed line)*

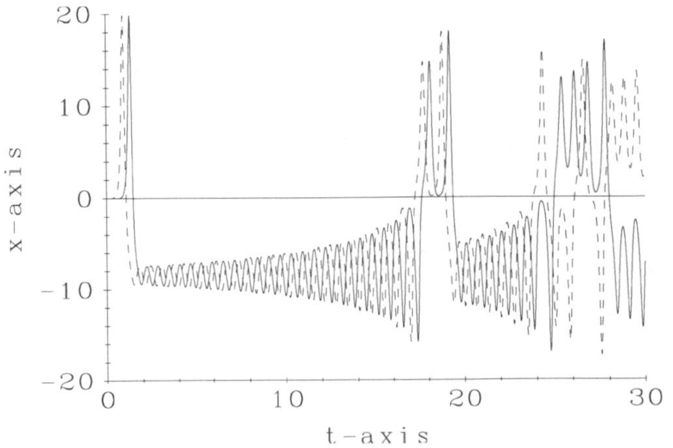

The ability to visualize surfaces in three-dimensional space is fundamental to the understanding of many of the types of derivatives and integrals in multivariable calculus. High school courses in solid geometry have disappeared. The tendency to skip comprehensive college algebra courses eliminates another way to learn solid analytic geometry. The task of learning even the basic list of algebraic formulas and their related surface shapes, such as the analytic geometry of the quadric surfaces, is now a primary objective of courses in multivariable calculus. One objective of visualization then is to give students a mental catalog of the common "household" formulas for space curves, surfaces, and solids, and their geometric representations. Another objective is to visualize the concepts of multivariable calculus. Images representing these concepts frequently involve the simultaneous presentation of multiple surfaces. To explain the differential, a tangent plane to the surface is required. Triple integrals are often described over regions between two surfaces. Flux integrals are described by representing a vector field on a surface. A successful environment for visualization would give students the capability to create and explore complex images such as those which appear in the textbooks of multivariable calculus.

To achieve these objectives, an integrated computer environment needs to have a varied graphics library of plot types together with symbolic algebra capabilities. As an indication of just how versatile a graphics environment should be, the problem of creating images of surfaces defined by three different algebraic modes is first discussed. Then, the problem of representing tangent planes is considered. It serves as an example of the problems of multi-surface plots and the usefulness of having symbolic differentiation in the environment.

Visualizing Surfaces

The surest way to understand the shape of a surface is to examine a physical model of the surface. Computer-generated images simulate that experience. As was indicated earlier, it is currently possible to generate a video simulation of a tour around the surface. Recall Bronowski's control of the image of the skull. Such simulations are effective because they provide many of the important kinetic cues to depth perception. For example, the kinetic cue of motion parallax indicates that an object further away appears to move a smaller distance laterally, than a closer object moving at the same speed. The dynamic rotation of an object allows the viewer to perceive the spatial organization of the surface of the object. Being able to control the image interactively allows the viewer to focus on some more complex aspect of the object which the viewer is trying to visualize. A complete discussion of depth perception can be found in *Psychology of Perception* [2].

Static images, on the other hand, must rely solely on the still cues for depth perception. When a wire-frame image with hidden line removal is generated, depth cues of perspective and interposition of surface elements are being used. In MATHLIB if the plot type "3-D-shaded" is used, depth cues of shading and shadow are added. *Mathematica* uses color images with shading based on three light sources. An additional cue called proximal brightness has been added. Integrated computer environments are beginning to provide choices for plotting images which have greater realism.

Figure 6. *Graph of*
$z = x^3 y - x y^3$

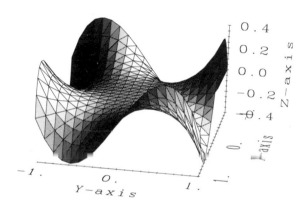

Surfaces are described in a variety of algebraic forms: as the graph of a function of two variables, $z = f(x,y)$, as level surfaces, $g(x,y,z) = C$, and as the images of parametric equations, $x = x(u,v)$, $y = y(u,v)$, $z = z(u,v)$. To be useful in multivariable calculus, the algebraic component of the integrated environment should accept all three of these algebraic formulations.

The plotting of the graph of a function of two variables on rectangles is usually quite routine. Figure 6 provides a plot of a function using the plot type "3-D-shaded" in MATHLIB.

Figure 7. *The level surface*
$w = 4x^2 + y^2 + z^2 = 144$

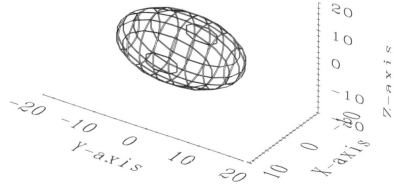

Using polar coordinates it is possible to present graphs on non-rectangular regions such as disks. But the image of a surface over an irregular region, such as all (x,y) for which $f_1(x) \leq y \leq f_2(x)$, and $a \leq x \leq b$, is needed to visualize double integrals yet is not commonly found in current software.

Surfaces are also described algebraically as the level surfaces of a function, $f(x,y,z) = C$, as is done in displaying equipotential surfaces. Display of such surfaces requires special numerical programs. Figure 7 shows the result of using the "3-D-level-surface" plot type in

Figure 8. *The cone*
x = ρ cos Θ,
y = ρ sin Θ,
z = ρ

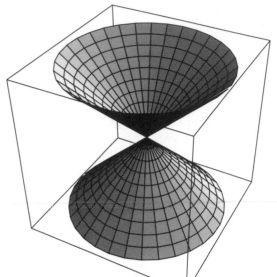

MATHLIB. The image is made up of bands that might be described as a "wicker-frame." The width of the bands can be controlled and can be thinned to create a wire-frame image.

Parametric surfaces arise naturally when describing the coordinate surfaces of systems of curvilinear coordinates such as with cylindrical and spherical coordinates. What would be most efficient is to have a single command for plotting a parametric surface. *Mathematica* has such a command called ParametricPlot3D. Figure 8 shows a plot of a conical coordinate surface from spherical coordinates using *Mathematica.*

In the absence of such a command, parametric plots can be built up by plotting successive coordinate curves of the surface in each parameter, but then hidden line removal is lacking.

In summary then, surfaces can be vizualized effectively if the integrated computer environment has plotting modes with different degrees of realism and will easily accept the algebraic descriptions of surfaces commonly used in multivariable calculus.

Visualization of Tangent Planes to Surfaces

In multivariable calculus, there are several important concepts related to differentiation which are visualized using images constructed from multiple surfaces. Partial derivatives and directional derivatives can be explained using a vertical plane through a selected point on the graph of a function, $z = f(x,y)$, and the tangent line through the point. The tangent plane to a surface utilizes different partial derivatives depending on the algebraic description of the surface as a graph, level surface or image of parametric functions, as seen earlier. Using these images of tangent planes, the differential can be explained as a local linear approximation to the function. The capability to plot multiple surfaces would have numerous applications to visualization in the differential calculus. Constructing an image of a tangent plane to the graph of a function, $z = f(x,y)$, then, is representative of a class of problems in the differential calculus.

Turning first to algebraic considerations, note that the capability for symbolic differentiation makes it easy to compute partial derivatives algebraically and therefore to code the equation for the tangent plane as it appears in textbooks. This emphasizes that what changes is the underlying function and the point of tangency.

Technical problems emerge for the graphics system. Two surfaces must be plotted in the same view volume. Superimposing one image on top of the other can produce confusion. In a sophisticated graphics system, the tangent plane might be added as a translucent panel. An attempt to emulate this using a "3-D-line plot" of the outline of a tangent rectangle is contained in Figure 9. The image was drawn using MATHLIB, which has a symbolic differentiator.

Once this is accomplished, a problem of perception emerges. Depending on the concavity of the surface, that tangent rectangle is either above or below

Figure 9. *Tangent plane to the graph of* **z = exp(-x² - y²)** *at (0.2, 0.2, 0.92)*

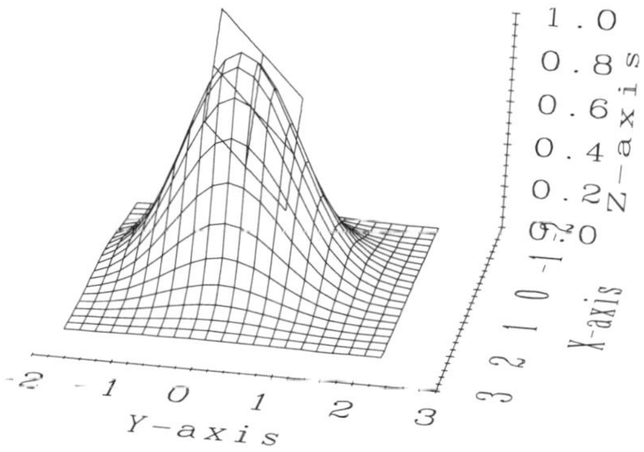

the surface of the function. Cues of depth perception become confused. The use of different colors for the two surfaces is an improvement and the use of fewer lines to define the underlying surface also helps. One comes to the realization that the effective textbook pictures of tangent planes to surfaces are done for carefully selected points on the surface and are viewed from advantageous eye postions.

Other solutions need to be explored. If the capability to continuously move the eye position about the image were available, it might be possible to determine a vantage point from which the tangency is understandable. In the realm of flight training simulators, images are constructed which realistically incorporate the kinetic cues of depth perception. That technology has not yet "trickled down" to the integrated environments the author has examined. The problem remains to be solved in future versions of these integrated environments.

The Role of an Integrated Environment in Teaching

The integrated environments described were originally created to meet the needs of research mathematicians and scientists. Here they are being evaluated for their usefulness in teaching. Integration of numerics, symbolic algebra and graphics is one of the most important advantages. Visualization is always available no matter what the project is. If the system is properly adapted to the needs of teaching, it may be possible for a student to use the system for beginning calculus through to professional research.

The graphics libraries of such systems enhance visualization by providing greater realism through their choice of plot types. Sometimes there is an excess of detail. TEMPLATE, part of MATHLIB, was constructed to do presentation graphics and contains controls for a dizzying array of parameters. Despite this richness, all the algebraic modes needed to construct surfaces in multivariable calculus cannot be coded by direct single commands. At this writing, MATHLIB (Version 11.3) lacks a single parametric surface command and *Mathematica* (Version 1.2) does not have a level surface command although it can sketch level curves. In addition, the capability to do multisurface plotting needed for multivariable calculus is missing from MATHLIB but is present in *Mathematica*. Perhaps as these needs are made known, the systems will be adapted to meet them in the future.

The symbolic algebra and numerical libraries of such systems add a new dimension to teaching. This capability also means that students must now be taught about the sources of error in algebraic and numerical procedures. In some cases, software utilizes well-known algorithms described in the literature, but without examining the code it is hard to assess the accuracy of the implementation. Sometimes documentation indicates the nature of the algorithms used as well as sources of code. At the first *Mathematica* conference, many users requested that documentation be expanded to provide greater information about algorithms. In the absence of such information, the task of determining the correctness of software is left to reviewers and users. It would also be helpful if professionals would develop test packages which could be used to evaluate the algorithms in software. The addition of such libraries is a definite advantage. One cost is the additional education required to use such tools, but this education is also a benefit.

Finally, the problems of choosing hardware to operate these large integrated packages should be mentioned. These are complex issues often dependent on many factors such as institutional budgets and previous purchases. At the University of the Pacific, MATHLIB resides on a VAX 785 which is networked to computer laboratories on campus. Students can access the program using the Versaterm emulator on Macintosh computers. Labs were constructed to serve either as an environment of personal computers or as a network system of graphics terminals attached to the VAX. MATHLIB was designed specifically to provide drivers for an extensive list of graphics terminals and can be configured flexibly. Apparently plans are in the works for configuring *Mathematica* in such networks. In general, efforts are underway to develop hardware systems which access the software through a network, with a view to reducing cost.

In their current state, integrated environments provide the computer algebra and graphics to meet a significant number of the needs for visualization in multivariable calculus. Their numerical and graphics libraries add a valuable dimension to differential equations. The graphics libraries of these systems should be enhanced to visualize the more complex images used in multivariable calculus.

The author wishes to thank Prof. Robert Borrelli of Harvey Mudd College, and Ned Freed, Dan Newman, and Kristin Hubner, all of Innosoft International, Inc., for their assistance in working with MATHLIB for classroom use.

Bibliography

[1]   Lorenz, Edward N., "Deterministic Nonperiodic Flow," *Journal of the Atmospheric Sciences,* 20(March, 1963), 130-141.
[2]   Dember, William N. and Joel S. Warm, *Psychology of Perception*, Second Edition, Holt, Rinehart and Winston, 1979.

Acknowledgements

The work leading to this article was supported in part by the National Science Foundation under grant number DMS-8851255.

Student-Generated Software for Differential Geometry

Thomas Banchoff and student assistants Jeff Achter, Rashid Ahmad, Cassidy Curtis, Curtis Hendrickson, Greg Siegle, and Matthew Stone

In a previous article [1], we described the progress made by a team of students working together with a professor to develop **Vector,** a software program for third semester calculus, to be used in an interactive laboratory environment. The original program enabled students to display graphs of space curves and functions of two variables, and subsequent improvements have added a number of features for displaying geometric objects associated with curves and surfaces. An enhanced version of the original program continues to be used as a supplementary tool in several courses in third-semester calculus at Brown University, with consultants on hand to assist students using the program and to receive suggestions for improving its effectiveness. At present we are using this program in an interactive laboratory associated with the introductory course in the differential geometry of curves and surfaces.

During the current semester, a new opportunity in the development of the program has arisen in the undergraduate differential geometry course. Two of the students in the project are enrolled in the course and are acting as laboratory assistants for a weekly hour-long session devoted to the specific topics of the course. The two dozen students in the class can return to the fifty-unit laboratory to run the program at other times when the laboratory is available to all students on a first-come first-serve basis. Students can work individually or together on assignments which require them to investigate the behavior of a curve or a surface or a family of such objects. Several of the challenges we have encountered are interesting in their own right, as we seek the best ways of utilizing the growing capabilities of the machines and the programs. By discussing these topics here, we wish to give some feeling for the way that this project is progressing as a true collaboration between students and instructors, and to show some of the ways our experience with the computer laboratory environment suggests changes in the choice of topics and the presentation of the subject matter.

The Cardioid Series

In dealing with parametric curves, it is often desirable to investigate not just one object but rather a family of curves. A particularly interesting example is the family which includes the cardioid. We define a family of polar coordinate function graphs depending on one parameter c, by
$$X(t) = ((c + \cos(t))\cos(t), (c + \cos(t))\sin(t)).$$
The value $c = 1$ (or $c = -1$) gives a cardioid with a cusp at the origin. What is the behavior of the other curves in the family?

Tom Banchoff has been a professor in mathematics at Brown University since 1967, specializing in the geometry of surfaces in 3- and 4-dimensional space.

The student assistants are mathematics-computer science majors at Brown University. Rashid Ahmad graduated in 1988 and is a student at the Columbia University School of Medicine. Greg Siegle is a junior and Jeff Achter, Cassidy Curtis, Curtis Hendrickson, and Matthew Stone are sophomores.

Using **Vector**, students can enter the coordinate functions for the above equation, then choose various values of c and display the corresponding curves, one at a time or several at once. An example is shown in Figure 1. One of the primary aims of the course is the analysis of singularity behavior, so it is especially important to analyze the curves near the critical position, say at $c = .9$ and $c = 1.1$. In the first case, the curve is locally convex with a double point at the origin; in the second, the curve is one-to-one, but it has a pair of inflection points and an interval where the curvature is negative. This behavior is typical for deformations of cusps, and students will recognize it again and again during the course.

■ **Figure 1**

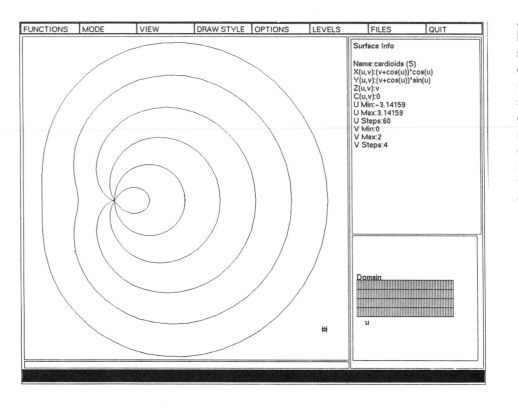

Are there any other interesting choices for c? A student entering various other values of c will easily note that there is a symmetry, and curves corresponding to opposite parameter values are congruent. This suggests a special treatment of the value $c = 0$, which gives a doubly-covered circle, unfolding into a curve with one double point when c is a small positive or a small negative number. What about large values of c? Will there always be inflection points if c is greater than 1? Computer investigation indicates that this is not so, and it seems that the curves are convex for all values of c greater than 2.

Naturally in a mathematics course it is not enough only to observe these phenomena; we also have to prove that what we observe is true. Experimentation with polar coordinate function graphs suggests a criterion for double points, either $r(t+\pi) = -r(t)$ for some t or $r(t) = 0$ for two different values of t. These conditions show that there will be double points when $|c| < 1$, as observed.

The existence of inflection points is equivalent to the vanishing of the numerator of the curvature
$$y''x' - x''y' = -rr'' + 2r'^2 + r^2$$
$$= -(c + \cos(t))(-\cos(t)) + \sin^2(t) + (c + \cos(t))^2$$
$$= 1 + c^2 + 3c\cos(t) + \cos^2(t) = 0.$$
This will have solutions exactly when $1 \le |c| \le 2$, as predicted by the images on the computer screen.

Animating Parameter Changes

In addition to keying in desired values of a parameter, a student can set up a sequence of examples by instructing the computer to change c from a beginning value to an end value in a certain number of steps. The images can then be played back to give an animated view of the deformation represented by the change of parameter values. This technique enables a student to "unfold" a singular phenomenon which arises naturally in the course of a one-parameter deformation.

Parallel Curves for Function Graphs

Animation techniques are especially effective when a deformation is related to a physical phenomenon, such as the propagation of wave fronts in the neighborhood of a curve. This phenomenon leads to the concept of *parallel curves*, where the parallel curve at distance r is obtained by moving r units along the unit normal at each point of the curve. Classical treatments of this subject emphasized the fact that if the distance is sufficiently small, the parallel curve to a smooth curve is smooth. With computer graphics, we can deal with a larger range of phenomena, and we can pay much more attention to the important subject of singularities of curves and families of curves.

A student can use **Vector** to enter in the equations for the coordinates of a parallel curve, with the parameter c as the distance to the parallel curve. Symbolically we may write $X_c(t) = X(t) + cU(t)$, where $U(t)$ is the rotation of the unit tangent vector $T(t) = X'(t)/|X'(t)|$ by $\pi/2$ radians. In early versions of **Vector**, students entered the explicit equation for each coordinate separately. The current version allows students just to enter the original curve and have the program calculate the auxiliary vectors automatically. By choosing various values of c, it is possible to see where the parallel curve develops cusps, and where the parallel curve intersects itself.

For the parabola $X_c(t) = (t, t^2)$, the student would enter the equations $x(t) = t - 2ct/\text{sqrt}(1 + 4t^2)$, $y(t) = t^2 + c/\text{sqrt}(1 + 4t^2)$, $-2 \leq t \leq 2$, and then select various values of c. For small values of c, the curve appears to be smooth, up to the value $c = 1/2$ where the curve seems to have a corner. Closer inspection shows that the curve is smooth although very flat at the point $X_{1/2}(0)$. For larger values of c, the curve has a pair of cusps and a single intersection point, a fact that is easy to check analytically.

In order to study the relationship between a curve and its parallel curve, it is important to display both of them at the same time and to relate them to the same coordinate system. When first given a curve, **Vector** computes the maximum and minimum x- and y-coordinates and displays the image on the largest possible square screen. It is possible to choose an option "View in Same Space" so that subsequent curves are referred to the coordinate system of the first curve, even if they do not fit on the screen. It is also possible to change the roles of curves so that the screen size is determined by any one of them. We can also resize the screen manually by selecting a square in which the visible screen will be redrawn, or zoom in on a particular section by selecting a square which is then expanded to fill the screen. At any stage in the investigation, it is possible to display the coordinate axes, or to choose a viewing space centered at the origin. The "bird's eye view" option shows the projection into the x-y-plane.

In general the variation of the vector $U(t)$ gives information about the way the curve deviates from a straight line, and we may define the *(geodesic) curvature* $k_g(t)$ by $U'(t) = -k_g(t)X'(t)$. It is then clear that the parallel curve is smooth, with $X'_c(t) = (1 - ck_g(t))X'(t)$, so the tangent lines at corresponding

Figure 2

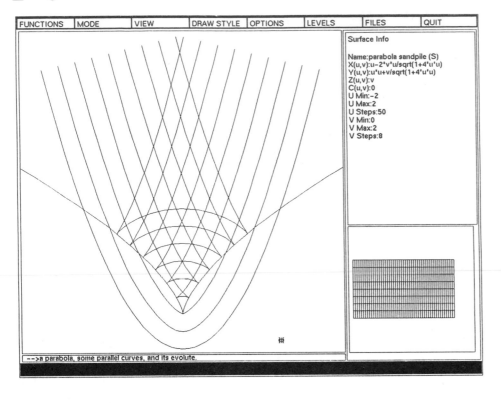

--->a parabola, some parallel curves, and its evolute.

points are the same and the direction of the unit tangent vector to the parallel curve is the same as that of the original if and only if $(1 - ck_g(t))$ is positive. The parallel curve has a singularity if the distance $c = 1/k_g(t)$, the *radius of curvature* of the curve at $X(t)$. The locus of singularities of the parallel curves is called the *focal curve* or *evolute* curve of X.

To illustrate this important fact, it is possible to display the curve and its evolute $E(t) = X(t) + (1/k_g(t))U(t)$ simultaneously, along with several parallel curves, as shown in Figure 2, or it is possible to set up an animation to show the cusps of parallel curves moving along the evolute.

Parallel Regions for Plane Curves

Since **Vector** was originally written to display surfaces, it is easy to use the two-parameter capability to create an entire family of parallel curves. We define a surface by $Y(u,v) = X(u) + vU(u)$, where $X(u)$ and $U(u)$ represent the curve and its unit normal vector in the plane, and where v goes from 0 to c in a certain number of steps. We always have the option of showing only the curves v = constant, and this family gives the desired set of parallel curves.

We get a bonus from this representation if we show instead the curves u = constant, along with the original curve. We then have the collection of normal lines to the curve, and the curve or points where nearby normals intersect is quite evident. We can then show that this singularity curve is the evolute of the original curve. Then $Y_u(u,v) = X'(u) + vU'(u) = (1 - vk_g(u))X'(u)$ and $Y_v(u,v) = U(u)$. Since $X'(u)$ and $U(u)$ are always linearly independent for a regular curve X, the only singularities occur if $v = 1/k_g(u)$, i.e. at the points of the evolute $E(u) = X(u) + (1/k_g(u))U(u)$.

At this stage, we can get additional information by using the color capabilities of the machine. We may assign colors to points of the curve according to the values of the parameter, and then color the points on the evolute similarly to establish the correspondence. This is especially clear when we use the same colors on the rays going out perpendicular to the curve.

Evolute Curves for Epicycloids

One of the most dramatic "discoveries" that appears from the use of this program is the evolute phenomenon for the family of epicycloids,

$X(t) = ((1 + c)\cos(t) - c\cos((1 + c)t/c), (1 + c)\sin(t) - c\sin((1 + c)t/c)).$

Entering in the equations of the curve together with its normal rays produced images which are unmistakably similar to the original curve, no matter what the value of *c* may be. Only if *c* is integral will the curve close off in the interval $0 \le t \le 2\pi$, but for other rational values of *c*, the curve will close an appropriate multiple of 2π, producing some striking images of curves and their (similar) evolutes. Of course it is then possible (for the better students) to prove the theorem that the evolute is indeed similar to the original curve.

Nesting of Osculating Circles of a Spiral

It is a fact that osculating circles of a spiral with monotonically increasing curvature are nested, i.e. the best approximating circle at the beginning of such an arc completely contains the corresponding circle at the endpoint. This result is surprising to many students. If a student draws a curve on a paper or blackboard and sketches in the osculating circles at two nearby points, the circles almost always appear to cross, even though the theorem predicts that they wil not intersect. **Vector** makes it possible to illustrate the theorem and to see what the collection of osculating circles really looks like. A student can enter a spiral arc, say the positive half of a parabola $X(u) = (u, u^2)$, $0 \le u \le 2$, along with a pair of its osculating circles, say the one at the origin, with radius 1/2 and center (0, 1/2) and the one at (1, 1) with center at (-4, 7/2) and radius $2/5^{3/2}$. The smaller circle is contained in the larger, and we don't need a computer to show that. But the computer can show even more. The program is set up to show surfaces, so it is possible to build an entire set of curves and to show as many of them as we wish. We can show not just two circles but the family of all osculating circles at all points of the parabola, and the visual evidence that they do not meet is compelling, as shown in Figure 3.

Figure 3

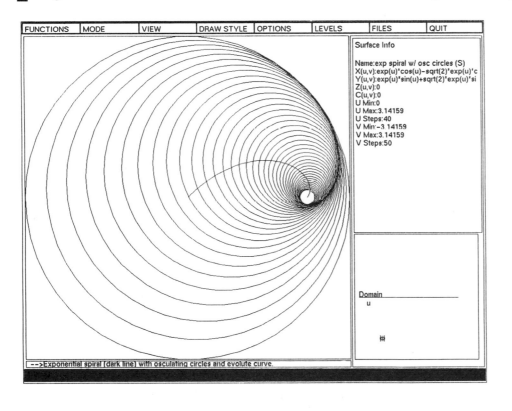

```
FUNCTIONS   MODE    VIEW    DRAW STYLE  OPTIONS   LEVELS    FILES         QUIT
```

Surface Info

Name:exp spiral w/ osc circles (S)
X(u,v):exp(u)*cos(u)-sqrt(2)*exp(u)*c
Y(u,v):exp(u)*sin(u)+sqrt(2)*exp(u)*si
Z(u,v):0
C(u,v):0
U Min:0
U Max:3.14159
U Steps:40
V Min:-3.14159
V Max:3.14159
V Steps:50

Domain
u

--->Exponential spiral (dark line) with osculating circles and evolute curve.

There are various capabilities of the machine that make these phenomena even clearer. We can color the points of the original curve according to the parameter value, then color the osculating circle at a particular point with the same color. We can also show a sequence of bands between successive osculating circles to make the nesting property even more evident.

Further Topics in the Geometry of Curves and Surfaces

Up to now we have discussed only the two-dimensional capabilities of the machine. We can gain even more insights by looking at objects in three-space. We first study surfaces associated with curves, such as the *tangential surface* $X(u) + vT(u)$, where $T(u)$ is the unit tangent vector of a space curve, or the *normal surface* $X(u) + vP(u)$, where $P(u)$ is the principal normal. We can then work with tube surfaces, like the *normal tube*

$$X(u) + r\cos(v)P(u) + r\sin(v)B(u)$$

which are the analogues of the parallel curves in the plane, in order to study the curvature and torsion properties of curves. We can define more general tubes, such as the *curvature tube*

$$X(u) + (1/k(u))(\cos(v)P(u) + \sin(v)B(u)),$$

a useful technique for modelling the growth of shells. A tangent surface is shown in Figure 4, and a normal tube is shown in Color Plate 3.

We can then go on to study surfaces in their own right, including parallel surfaces and focal surfaces of function graphs and of parametric surfaces. All of these capabilities are already available on **Vector** once we enter the appropriate combinations of functions. At the same time, it is clear that these more complicated topics are stretching the capabilities of the program, and we can anticipate a further redesign in the future. We look forward to the ability to work with vector functions, and to calculate quantities like curvatures and principal directions, without having to enter the explicit equations for each coordinate. (This capability existed to a certain extent in the program EDGE developed for an earlier version of the interactive laboratory environment in differential geometry, as described in [2].)

■ **Figure 4**

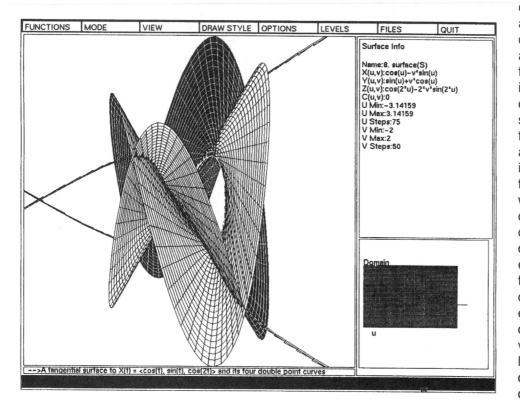

We have learned a great deal in the development of these programs, using the ideas of differential geometry to suggest new directions for the project in the calculus of curves and surfaces. As mentioned in previous articles, the interaction between faculty and students in this collaboration is one of the most rewarding aspect of the whole enterprise. We look forward to the next phases of our project, and the continued use of interactive software in geometric investigations.

References

[1]   Banchoff, Thomas, and Student Associates, "Student-Generated Interactive Software for Calculus of Surfaces in a Workstation Laboratory," *UME Trends*, (August 1989), 7-8.

[2]   Banchoff, Thomas and Richard Schwartz, "EDGE: The Educational Differential Geometry Environment," in *Educational Computing in Mathematics*, T. F. Banchoff, et.al., (eds), Elsevier Science Publishers B.V., North Holland, 1988, 11-30.

Viewing Some Concepts and Applications in Linear Algebra

Thomas Hern and Cliff Long

We present several examples of visualization techniques for use in teaching linear algebra and its applications. We begin with the fundamental notions of linear transformation, determinant, eigenvalue, and eigenvector; and then move on to topics that have not been included in traditional linear algebra courses until recently: least squares, singular value, pseudoinverse, ill-conditioning, and the convergence of regular Markov chains. These later topics address ideas in applications, statistics, and computation that are becoming increasingly important. We also give samples of computations with MATLAB™ software.

Fortunately three dimensions is adequate for understanding many of these ideas, so that pictures such as those presented here will be quite useful. Three dimensions is also often necessary to allow for the complexity that two dimensions may not be able to show. The common theme in each example will be to view the effects of a linear transformation on a basic shape in $\Re^3$: a cube, ball, or tetrahedron. (Occasionally, for clarity, we resort to analogous shapes in $\Re^2$.) The choice of the basic shape will be influenced by the application and what is to be illustrated. Although we are primarily addressing teachers of linear algebra, we will still present basic definitions and ideas for completeness. There should be something for anyone interested in linear algebra and its applications.

The Basic Unit Cube and Its Image Parallelepiped

Since each vector $x = (a,b,c)$ in $\Re^3$ can be written as a linear combination $ae_1+be_2+ce_3$ of the standard basis vectors $e_1 = (1,0,0)$, $e_2 = (0,1,0)$ and $e_3 = (0,0,1)$, we know the image of any vector x under a linear transformation L if we know the images of the e_i. The geometry in $\Re^3$ implied by the properties of a linear transformation ($L(x_1+x_2) = L(x_1) + L(x_2)$ and $L(kx) = kL(x)$) is very powerful. We wish to study this geometry relative to the *basic unit cube*, shown in Figure 1, which has one vertex at the origin $0 = (0,0,0)$ and end points of adjacent edges given by the standard basis vectors e_1, e_2 and e_3.

The linear transformation L may be written in matrix form as $L(x) = Ax$, where the columns of the matrix A are the image vectors $c_i = L(e_i)$. The origin maps into the origin, since $L(0) = 0$, and the other vertices of the cube such as $v_4 = e_1 + e_2$ map into $w_4 = L(v_4) = L(e_1 + e_2) = L(e_1) + L(e_2) = c_1 + c_2$, etc. So in general the vertices of the basic cube are transformed into the vertices of a parallelepiped with a vertex at the origin. The edges meeting at 0 are the column vectors of A. In Figure 1a we have a general example with $c_1 = (1,1,2)$, $c_2 = (0,2,1)$, and $c_3 = (-1,2,2)$.

Thomas Hern is Associate Professor of Mathematics and Statistics at Bowling Green State University in Ohio. His work has been in probability theory and computer graphics. His current interests are in the mathematics of computer graphics and computer-aided design and their use in the teaching of mathematical ideas.

Cliff Long is Professor of Mathematics and Statistics at Bowling Green State University, and is currently on the MAA Board of Governors. He participated in an NSF Institute on Computer Graphics for Learning Mathematics at Carleton College in 1972-74, and continues this interest. His current work is in computer-aided geometric design and numerically-controlled milling.

Any point x on the edge from vertex v_i to vertex v_j can be written as
$$x = (1-t)v_i + tv_j = v_i + t(v_j - v_i), \text{ with } 0 \le t \le 1,$$
and the parameter t represents the position along the segment from v_i to v_j. Then $L(x) = L(v_i) + t(L(v_j) - L(v_i)) = w_i + t(w_j - w_i)$, which is in the same position on the line segment joining these image vertices. Hence the image of the edge of the cube between v_i and v_j is the edge between the corresponding vertices of the image parallelepiped. Midpoints of edges of the basic cube also map into midpoints of their image edges, and similar arguments show that faces and interior points of the basic cube map into corresponding faces and interior points of the image parallelepiped.

The columns of the matrix A form a basis (at least in the case of full rank) for the column space of A. If in addition the columns are orthogonal, then the image parallelepiped is a rectangular box, as shown by the example in Figure 1b. If the columns are unit vectors, the image parallelepiped is a cube of volume one. The matrix is then called *orthogonal* (better perhaps "orthonormal") and the image of the basic cube is again a cube. An important property of orthogonal matrices, which we will use extensively later, is that they preserve length and angle, and thus shape.

Under a linear transformation, the shape of the image of the unit cube is the same as the shape of any translate of the unit cube. (A translate of the cube by a vector d maps each x in the cube to image $L(x + d)$ given by $L(x)+L(d)$, which is a translate of its image by the vector $L(d)$.)

In contrast to linear transformations, we see in Figure 1d the image of the unit cube under a nonlinear transformation from $\Re^3$ to $\Re^3$, which is no longer even a polyhedron. In addition, the image of a translate of the basic cube may look nothing at all like the image shown in Figure 1d. The image volume changes with location as well.

The Determinant as the Volume of the Image Parallelepiped

A fact that is often not emphasized is that the volume of the image parallelepiped under $L(x) = Ax$ is equal to $|\det(A)|$. (The proof given by Strang [13, pp. 212, 234] applies, since edges adjacent to 0 are the columns of A and the rows of A^T which has the same determinant as A.) We can see this in Figure 1b: the determinant is 2 and the image volume is twice that of the unit cube. In the case of an orthogonal matrix the determinant is ± 1, which is consistent with the fact that the image is a cube of volume one. If the whole matrix is multiplied by 2, thus doubling all the columns, then the image parallelepiped has each edge doubled and its corresponding volume multiplied by 2^3. This makes quite plausible the general property that $\det(kA) = k^n \det(A)$.

Singular and Nearly Singular Transformations

This volume interpretation leads us to the important point that a singular matrix A, which has $\det(A) = 0$, has a degenerate image parallelepiped with zero volume. The three columns lie in a plane, since they are linearly dependent, and hence the basic cube is flattened into a plane. Figure 1c shows this case where the third column of the matrix for Figure 1a is changed to the average of the first two, so that the new matrix is of rank 2 (i.e., 2 columns are linearly independent, but not all 3). The image parallelepiped is in a plane and has positive area but zero volume. In general when the rank is 1, the image parallelepiped is a line segment, and when the rank is 0 the image of the basic cube is the origin itself. These ideas are emphasized in Green [5, p. 172] and were illustrated by Long [9] in an animated super-8 film.

■ Figure 1

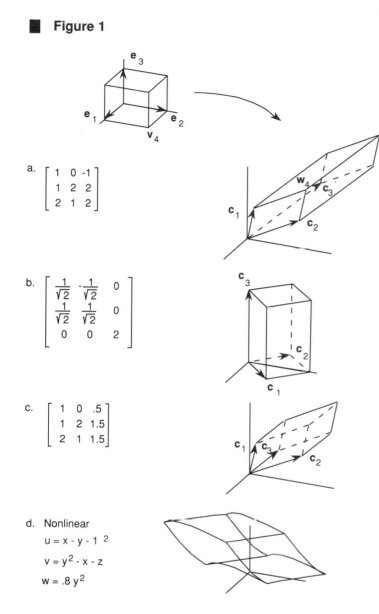

a. $\begin{bmatrix} 1 & 0 & -1 \\ 1 & 2 & 2 \\ 2 & 1 & 2 \end{bmatrix}$

b. $\begin{bmatrix} \frac{1}{\sqrt{2}} & -\frac{1}{\sqrt{2}} & 0 \\ \frac{1}{\sqrt{2}} & \frac{1}{\sqrt{2}} & 0 \\ 0 & 0 & 2 \end{bmatrix}$

c. $\begin{bmatrix} 1 & 0 & .5 \\ 1 & 2 & 1.5 \\ 2 & 1 & 1.5 \end{bmatrix}$

d. Nonlinear
$u = x - y - 1^2$
$v = y^2 - x - z$
$w = .8\, y^2$

If the matrix A is nearly singular, the image parallelepiped is nearly planar. What often happens in practice is that a matrix may be singular, but roundoff error due to the floating point arithmetic of computers makes it appear to be nonsingular; or the matrix may be nearly singular itself, such as often happens with least squares in regression problems in statistics. There are practical computational difficulties in both instances. We will come back to this important case when we consider ill-conditioned matrices.

If the determinant is not zero, then the transformation is nonsingular and one-to-one, the inverse is well defined, and a unique solution for $Ax = b$ exists. In this case the image of the basic cube does not collapse to a plane figure, and the image parallelepiped under the inverse transformation maps back onto the original cube. From the volumes, we can see that the determinant of the inverse of a nonsingular matrix A has to be $1/\det(A)$.

The Unit Ball and its Image Ellipsoid

We have seen in the above examples that some vectors have their lengths $\|x\|$ changed under a transformation, and others do not. Of course if an image vector Ax is larger than x, i.e., $\|Ax\| > \|x\|$, then the image Ax has its length decreased under A^{-1}. We are interested in those vectors whose images are changed the most (and the least) in length. The unit cube depends too much on the choice of basis vectors, and tends to cover up these other things that may be going on. In an attempt to find a basic shape which may detect this change of length, we consider the *unit ball*, i.e., all x such that $\|x\| \le 1$. The image of the unit ball is an ellipsoid (and its interior). This may seem intuitively clear, and is often stated without justification. Determining the directions and the lengths of the axes of the ellipse is of special interest in numerical linear algebra.

Real Symmetric Matrices

If A is real and symmetric, the geometry drawn in Figure 2 is fairly simple: we can show that the axes of the image ellipsoid agree with the *eigenvectors* of the matrix, i.e., nonzero vectors x for which $Ax = \lambda x$ for some λ. The number λ is the associated *eigenvalue* for x. For a real symmetric matrix, there is a complete set of orthogonal eigenvectors, with corresponding real eigenvalues [13, p. 309]. A standard result in linear algebra courses is that a symmetric matrix A can then be *diagonalized* in a special way, i.e., represented in the form $A = Q\Lambda Q^T$ where Q is an orthogonal matrix with columns made up of unit eigenvectors x_i of A, and Λ is a diagonal matrix with the eigenvalues λ_i of A on the diagonal.

For the symmetric matrix $\qquad A = \begin{bmatrix} 1.25 & -.75 \\ -.75 & 1.25 \end{bmatrix}$

the unit eigenvectors are $x_1 = (1/\sqrt{2})(1,1)$ and $x_2 = (1/\sqrt{2})(-1,1)$, with corresponding eigenvalues $\lambda_i = 0.5$ and 2, respectively. These are drawn in Figure 2. Since Q takes the basis vectors e_i to the eigenvectors (its columns), the transformation Q^T, which is the same as Q^{-1}, rotates (and perhaps reflects) these eigenvectors to the coordinate axes. The diagonal matrix Λ then stretches them by the eigenvalues λ_i. So the unit circle (since we are in $\Re^2$) is stretched into an ellipse whose semi-axis lengths are $|\lambda_i|$. Finally Q rotates the result back to the original eigenvector orientation. The orthogonal matrices Q and Q^T preserve the shape, so the only shape change comes from the stretching of the circle by the diagonal matrix Λ into an ellipse. The situation in $\Re^3$ is similar.

The length of the longest image of a unit vector under A is referred to as the *norm* of A, denoted $\|A\| = \max \{\|Ax\| : \|x\| = 1\}$. (From the properties of a linear transformation it follows that this is the same as $\max \|Ax\|/\|x\|$ for all x.) Thus the norm of a real symmetric matrix is $\|A\| = $ largest $|\lambda_i|$, which is 2 in the example of Figure 2.

Figure 2: *Diagonalization of a real symmetric matrix: $A=Q\Lambda Q^T$*

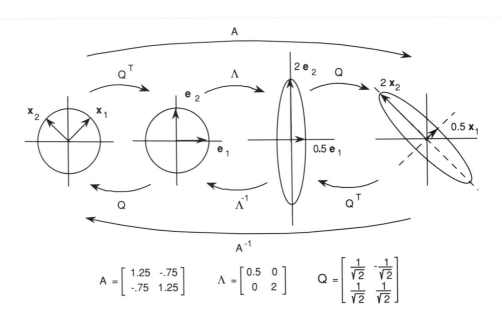

$$A = \begin{bmatrix} 1.25 & -.75 \\ -.75 & 1.25 \end{bmatrix} \quad \Lambda = \begin{bmatrix} 0.5 & 0 \\ 0 & 2 \end{bmatrix} \quad Q = \begin{bmatrix} \frac{1}{\sqrt{2}} & \frac{-1}{\sqrt{2}} \\ \frac{1}{\sqrt{2}} & \frac{1}{\sqrt{2}} \end{bmatrix}$$

We can also observe the action of A^{-1} in Figure 2, where each point of the image ellipse maps back to a point of the unit circle. The "eigendirections" for A and A^{-1} agree, and the *dominant eigenvector* of A (which is the eigenvector with the largest absolute eigenvalue) is the eigenvector of A^{-1} with the smallest eigenvalue. Whatever A does, A^{-1} must undo. If A stretches a vector by factor λ, then A^{-1} must shrink it by that same factor (i.e., stretch it by $1/\lambda$). Hence in the example above, $\|A^{-1}\| = 1/(\text{smallest } |\lambda_j|) = 1/0.5 = 2$.

General Matrices and the Image Ellipsoid

The eigenvectors for a matrix need not line up with the axes of the image ellipsoid. This is easily seen in Figure 3 for the transformation in $\Re^2$ with matrix

$$A = \begin{bmatrix} 1 & 1.5 \\ 0 & 1.5 \end{bmatrix}$$

Since this is a triangular matrix, its eigenvalues are the diagonals $\lambda_1 = 1$ and $\lambda_2 = 1.5$. Corresponding unit eigenvectors are

$$x_1 = (1,0) \text{ and } x_2 = 1/\sqrt{10}(3,1) = (.9487, .3162).$$

(Approximate decimal calculations here and later were done with MATLAB. For a sample session, see Figure 15.) These vectors are shown in Figure 3 with $A\boldsymbol{x}_1 = \boldsymbol{x}_1$ and $A\boldsymbol{x}_2 = 1.5\,\boldsymbol{x}_2$ in the image. The image of the unit circle, $\|\boldsymbol{x}\| = 1$, under A turns out to be the ellipse on the right-hand side of Figure 3. The axes of this ellipse are certainly not in the eigenvector directions, since the eigenvectors are not orthogonal. We also see that the norm $\|A\|$ of the matrix in this example, the largest $\|A\boldsymbol{x}\|$ for which $\|\boldsymbol{x}\| = 1$, is clearly larger than the largest eigenvalue.

One way to see that this image is an ellipse is to write the transformation in terms of elementary matrices:

$$A = \begin{bmatrix} 1 & 1.5 \\ 0 & 1.5 \end{bmatrix} = \begin{bmatrix} 1 & 1 \\ 0 & 1 \end{bmatrix}\begin{bmatrix} 1 & 0 \\ 0 & 1.5 \end{bmatrix} = A_2 A_1$$

Then we can think of A as a composition of A_1, a stretching of 1.5 in the $\boldsymbol{e}_2$ direction, followed by a shear A_2, namely $\boldsymbol{e}_1$ goes to $\boldsymbol{e}_1$ and $\boldsymbol{e}_2$ goes to $\boldsymbol{e}_2 + \boldsymbol{e}_1$. We have drawn these in Figure 3. The top of the image ellipse under A_1, (0,1.5), is sheared to (1.5,1.5) by A_2. This is the highest point of the final ellipse, and not the end of the axis. From this we can see that the slope of the major axis is less than one. But this approach doesn't tell how to find the axes of the ellipse and their lengths.

Before proceeding, let us briefly consider the frequently overlooked geometry of the transpose A^T shown in Figure 4. It will be especially useful later in relating the geometry of A^{-1} to that of the pseudoinverse A^+. Comparing Figures 3 and 4 shows that while the size and shape of the image under A^T appear to be (and indeed are) the same as that of A, the directions of maximum stretching and shrinking of the image under A^T are not the same as for A (and thus A^{-1}). The eigenvectors of A^T are different than those of A, but the eigenvalues are the same.

Figure 3: *The image ellipse of the unit circle*

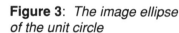

A

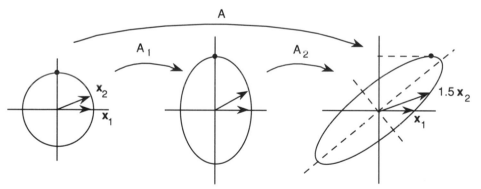

Figure 4

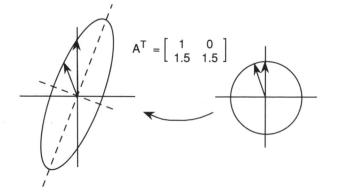

$$A^T = \begin{bmatrix} 1 & 0 \\ 1.5 & 1.5 \end{bmatrix}$$

Using Quadratic Forms to Find the Image Ellipsoid and Norm

To find the value of the norm of a nonsingular matrix A and the image ellipsoid, we use a method similar to [2]. To find the image of the unit sphere, consider $\boldsymbol{v} = A\boldsymbol{u}$ for $\|\boldsymbol{u}\| = 1$. Then $\|A^{-1}\boldsymbol{v}\| = \|\boldsymbol{u}\| = 1$. From this we get

$$1 = \|A^{-1}\boldsymbol{v}\|^2 = (A^{-1}\boldsymbol{v})^T A^{-1}\boldsymbol{v}$$
$$= \boldsymbol{v}^T(A^{-1})^T A^{-1}\boldsymbol{v} = \boldsymbol{v}^T B\boldsymbol{v}.$$

The symmetric matrix $B = (A^{-1})^T A^{-1}$ is positive definite (as is any matrix of the form $C^T C$, where C is invertible), and so B has positive eigenvalues and its complete set of eigenvectors can be chosen to be orthonormal. Hence the diagonalization of B gives a sum of squares, with the coefficients of the terms being the positive eigenvalues of B, and thus $v^TBv = 1$ is an ellipsoid. (See page 359 of [1].)

Moreover this gives us the length of the semi-major axis as the largest reciprocal of the square root of an eigenvalue of B and its direction as the associated eigenvector. But $B = (AA^T)^{-1}$, and so B has the same eigenvectors as AA^T and has reciprocal eigenvalues. Thus if AA^T has largest eigenvalue $\mu_1 > 0$, with unit eigenvector v_1, then the semi-major axis of the image ellipse is $\sqrt{\mu_1}v_1$.

In the example of Figure 3 above, the eigenvalues of AA^T are $\mu_1 = .4451$, with unit eigenvector $v_1 = (.6257, -.7800)$, and $\mu_2 = 5.0549$, with $v_2 = (.7800, .6257)$. This gives us the ellipse of Figure 3. Since $\sqrt{\mu_1} = .6672$, and $\sqrt{\mu_2} = 2.2483$, the semi-major axis is $\sqrt{\mu_2}v_2 = (1.7537, 1.4048)$ (with slope less than 1 as predicted), the minor axis is $\sqrt{\mu_1}v_1 = (.4174, -.5204)$, and the norm, $\|A\| = \sqrt{\mu_2} = 2.2483$. These are drawn on the right in Figure 5.

Figure 5: $A = Q_1\Sigma Q_2{}^T$

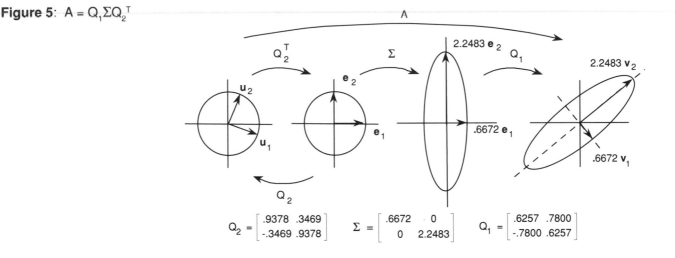

$$Q_2 = \begin{bmatrix} .9378 & .3469 \\ -.3469 & .9378 \end{bmatrix} \qquad \Sigma = \begin{bmatrix} .6672 & 0 \\ 0 & 2.2483 \end{bmatrix} \qquad Q_1 = \begin{bmatrix} .6257 & .7800 \\ -.7800 & .6257 \end{bmatrix}$$

What unit vectors in the example are mapped into the v_i? These are clearly $u_i = A^{-1}v_i$, drawn on the left in Figure 5, which are $u_1 = (.9378, -.3469)$ and $u_2 = (.3469, .9378)$. Note that these are orthogonal: if we apply the same argument to A^{-1} as we did to A, v_1 is stretched the most to u_1, and v_2 is stretched the least to u_2, and so u_1 and u_2 are eigenvectors of the symmetric matrix A^TA.

Singular Values and the Image Ellipsoid

The situation for the general nonsingular matrix in Figure 5 looks very much like the diagonalization in Figure 2 for the symmetric matrix $A = Q\Lambda Q^T$, where Q was an orthogonal matrix with columns made up of unit eigenvectors x_i of A, and Λ a diagonal matrix with the eigenvalues λ_i of A on the diagonal.

We form orthogonal matrices Q_1 with the v_i as columns and Q_2 with the u_i as columns, and let Σ be a diagonal matrix with the $\sigma_i = \sqrt{\mu_i}$ on the diagonal. Looking at Figure 5, we see that $Q_2{}^T = Q_2{}^{-1}$ rotates the vectors u_1 and u_2 to the base vectors e_1 and e_2, then Σ stretches by the factors $\sigma_i = \sqrt{\mu_i}$ which are the lengths of the axes (the positive square roots of the nonzero eigenvalues of A^TA), then Q_1 rotates these to multiples of v_1 and v_2. Thus we can write A as $Q_1\Sigma Q_2{}^T$. The entries σ_i of Σ are called the *singular values* of A, and $A = Q_1\Sigma Q_2{}^T$ is called the *singular value decomposition* of A. For the example above, these matrices are shown in Figure 5.

This decomposition can be performed even if A is singular or not square. Some of the μ_i may be zero, and Q_1 and Q_2 may be different sizes, but this can always be done (see Strang [13, p 443], Long [10], and Blank, Krikorian, and Spring [2].)

THEOREM: (Singular Value Decomposition) *Any m x n matrix A can be factored into $A = Q_1\Sigma Q_2{}^T$. The columns of the m x m orthogonal matrix Q_1 are eigenvectors of AA^T, and the columns of the n x n orthogonal matrix Q_2 are eigenvectors of A^TA. Σ is an m x n matrix with entries on the diagonal which are the singular values of A, and zero elsewhere.*

Although not yet taught in most linear algebra courses, the singular value decomposition is extremely important: "The SVD has become fundamental in scientific computing" [13, p. 197], "In any introductory course on matrices it deserves a place near the center." [7, p. 413] "The most reliable method for computing the coefficients for general least square problems ..." [4, p. 195].

What we see then is that the fundamental geometry needed for determining the norm and image ellipsoid of a general matrix A is connected with the eigenvectors of the matrices A^TA and AA^T. In the special case that A is a real symmetric matrix, these vectors are also eigenvectors of A (which will be seen later in Figure 6).

An Eigencube and Its Image

While the image under A of the unit ball is important for the norm of A, it does not automatically give us useful information about the eigenvectors of A, and as we have seen, may even be misleading. The above discussion suggests the use of a different basic shape for representing the transformation of a symmetric matrix: a cube with the edges being unit eigenvectors of the matrix. Let's call it an *eigencube*. (When the matrix is symmetric we have a full set of orthogonal eigenvectors.) For a general matrix A, it is the symmetric matrix A^TA and its eigencube which are crucial both for the norm and the pseudoinverse A^+. We first consider the symmetric case as a lead-in to the general.

Symmetric Matrix A and the Eigencube of A

In Figure 6, we show an eigencube for a nonsingular symmetric matrix $A = Q\Lambda Q^T$, along with its image under this transformation. For this A, we have $A^T = (Q\Lambda Q^T)^T = Q\Lambda^TQ^T = Q\Lambda Q^T = A$. If we now apply the transformation A^T to the last rectangular image in the top row of Figure 6, we get the geometry shown in the bottom row. The image of the eigencube of A under the combined transformation A^TA is just a stretched version of this same eigencube, with the stretching factors being the squares of the eigenvalues of A. Thus the eigenvectors of A are also eigenvectors of $A^TA = A^2$ with corresponding eigenvalues λ_i^2.

General Matrix A and the Eigencube of A^TA

A real matrix A may not be so nice as to have a complete set of orthogonal eigenvectors and real eigenvalues to use in the geometry of Figure 6. But when we follow A by A^T to obtain A^TA, then this new matrix is symmetric, positive semi-definite, and as noted before, it then has a complete set of orthonormal eigenvectors u_i, and nonnegative eigenvalues. The square roots σ_i of these eigenvalues are the singular values of A, and will serve as useful substitutes for the missing eigenvalues of the matrix A.

Figure 6: *Action of A and $A^T A$ on eigencube of a nonsingular real symmetric matrix A*

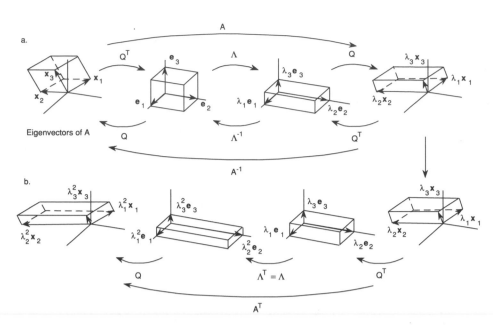

This leads to a direct generalization of the geometry of Figure 6 for a non-symmetric matrix A using the singular value decomposition, $A = Q_1 \Sigma Q_2^T$. We replace the nonexistent eigencube of A by the eigencube of $A^T A$, and the orthogonal matrices Q and Q^T by the corresponding orthogonal matrices Q_1 and Q_2^T. Recall that the columns of Q_1 and Q_2 are the unit eigenvectors of AA^T and $A^T A$ respectively. In a way similar to what we did in Figure 5, we first rotate this eigencube by Q_2^T so that it lines up with the coordinate axes. Then we stretch this new cube into a box by the diagonal matrix Σ of singular values of A. Finally, we rotate this new box by the orthogonal matrix Q_1 whose columns are eigenvectors of the symmetric matrix AA^T. The geometry related to this situation, shown in Figure 7, is directly connected with the singular value decomposition.

This form

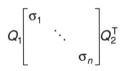

implies that the norm $\|A\|$, the maximum that any unit vector is stretched by A, is equal to the largest σ_i, since the orthogonal matrices Q_1 and Q_2^T preserve length. One vector which is stretched this amount is the dominant unit eigenvector of $A^T A$, as seen in the top row of Figure 7. When A is real symmetric, $\sigma_i = |\lambda_i|$, and so in that case we have $\|A\| =$ largest $|\lambda_i|$.

$A^T = Q_2 \Sigma Q_1^T$ is illustrated in the bottom row of Figure 7 and is the generalization of Figure 6. The combination $A^T A$ of Figure 7 clearly indicates the role of the eigenvectors $\boldsymbol{u}_i$, $\boldsymbol{v}_i$ and singular values σ_i in the transformation A. A takes unit eigenvectors $\boldsymbol{u}_i$ of $A^T A$ into eigenvectors $\sigma_i \boldsymbol{v}_i$ of AA^T, and A^T takes eigenvectors $\sigma_i \boldsymbol{v}_i$ of AA^T into eigenvectors $\sigma_i^2 \boldsymbol{u}_i$ of $A^T A$.

The Inverse and Pseudoinverse

If A^{-1} exists, then from the SVD we get $A^{-1} = (Q_2 \Sigma Q_1^T)^{-1} = Q_2 \Sigma^{-1} Q_1^T$, with Σ^{-1} having diagonal elements $1/\sigma_i$. Thus $\|A^{-1}\| = 1/$(smallest σ_i). It is clear from

Figure 7: *Action of nonsingular A and A^TA on eigencube of A^TA*

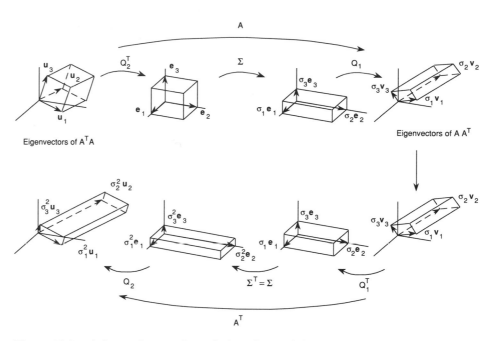

Figure 7 that information on the relative sizes of the largest and smallest values of σ_i is obtained from the shape of the image of the eigencube of A^TA.

However, if A is singular, Σ^{-1} does not exist, and we define a new matrix: $\Sigma^+ = \mathrm{diag}(d_i)$, with $d_i = 1/\sigma_i$ when $\sigma_i \neq 0$, and $d_i = 0$ if $\sigma_i = 0$. The corresponding matrix $A^+ = Q_2 \Sigma^+ Q_1{}^T$ is called the *pseudoinverse* of A.

Projection onto the Row and Column Spaces

Figure 8 shows the action of the matrix A^+A in the singular case. The matrix A maps all of $\Re^3$ onto the column space of A. Since $\boldsymbol{u}_i$ is mapped to $\sigma_i \boldsymbol{v}_i$, either $\boldsymbol{v}_i$ is in the column space of A or $\sigma_i = 0$. When $\sigma_i = 0$, $\boldsymbol{v}_i$ is mapped back to $\sigma_i \boldsymbol{u}_i = \boldsymbol{0}$ under both A^+ and A^T, so it is in the null space of A^T (called the *left null space* of A). The $\boldsymbol{v}_i$ with nonzero σ_i form a basis for the column space of A, since they are adequate in number and perpendicular to the left null space. A^+

Figure 8: *Singular A and pseudoinverse A^+*

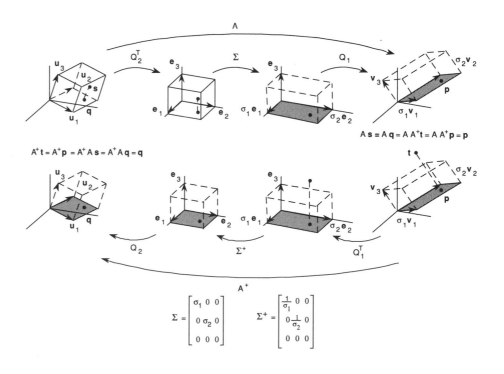

$$\Sigma = \begin{bmatrix} \sigma_1 & 0 & 0 \\ 0 & \sigma_2 & 0 \\ 0 & 0 & 0 \end{bmatrix} \qquad \Sigma^+ = \begin{bmatrix} \frac{1}{\sigma_1} & 0 & 0 \\ 0 & \frac{1}{\sigma_2} & 0 \\ 0 & 0 & 0 \end{bmatrix}$$

maps all of $\Re^3$ onto the row space of A, which is spanned by the subset of the eigenvectors $\boldsymbol{u}_i$ of $A^T A$ which have nonzero eigenvalues. The singular value decomposition provides an orthonormal basis for the column space and for the row space of the matrix A.

It is apparent from Figure 8 that if we restrict the action of A to the row space of A, then A^+ is really the inverse of A for this domain. As a matter of fact, for any matrix A, A^+ really is opposite and parallel in structure to A. The action of A can be thought of as first projecting to the row space and then mapping one-to-one onto the column space, while the action of A^+ projects to the column space and maps one-to-one back to the row space of A. Figure 8 illustrates this parallel form for a matrix A: $A\boldsymbol{s} = \boldsymbol{p}$ and $A^+\boldsymbol{t} = \boldsymbol{q}$.

Least Squares and Numerical Considerations

In applied problems we are frequently asked to obtain a best linear fit to a set of data points. This leads to having to solve a system of linear equations which has no solution. The least squares process leads to an approximate or "pseudo" solution to this linear system. Unfortunately, many such problems involve matrices which are "bad" from a numerical standpoint. These are matrices for which small numerical errors in the data (including computer roundoff) often cause large changes in the solution, making it essentially worthless. Such matrices must be identified. We will consider this problem before proceeding to the least squares problem, and will utilize the singular value decomposition once again.

Ill-Conditioned Matrices

Recall that the image parallelepiped of a singular 3x3 matrix is at best planar (since the determinant is zero and hence the volume of the image is zero). A matrix is "nearly singular," or ill-conditioned, when the corresponding image parallelepiped is flattened out a lot — almost planar. By that we mean that it is stretched considerably in one direction relative to the stretch or shrink in another direction. For example, the matrix

$$A = \begin{bmatrix} 1 & 0 & 0 \\ 0 & 100 & 0 \\ 0 & 0 & .02 \end{bmatrix}$$

has a maximum expansion factor of 100 (in the x_2 direction) and a minimum expansion factor of .02 (in the x_3 direction). The *condition number* of the matrix, cond(A), is given by the ratio of the maximum expansion factor of A to its minimum expansion factor.

To illustrate why matrices with large condition numbers, which are referred to as *ill-conditioned matrices*, may be troublesome, let's look at the transformation defined by the matrix A above as simulated in Figure 9 (going back to the unit ball), and relate it to solving the linear system $A\boldsymbol{x} = \boldsymbol{b}$. (Actually if the scales were the same on the coordinate axes, then the image ellipsoid would be even longer and skinnier than shown here.) With the matrix A above and $\boldsymbol{b} = (0, 80, 0)$, we obtain the solution $\boldsymbol{x} = (0, .8, 0)$. However, if we have a slight error in the problem, and use $\boldsymbol{b} = (0, 80, .01)$, the corresponding solution is $\boldsymbol{x} = (0, .8, .5)$. The relative error in the problem, $\|\Delta\boldsymbol{b}\| / \|\boldsymbol{b}\| = .01/80 = .00125$, is quite small. However, the relative error in the solution, $\|\Delta\boldsymbol{x}\|/\|\boldsymbol{x}\| = .5/.8 = .625$, is quite large in comparison (0.125 % error in the problem yields a 62.5% error in the solution.) So, small errors in the problem can be magnified considerably in the solution, making it unstable numerically.

In some cases this numerical sensitivity would result in inaccurate answers no matter what algorithm might be used to solve the system. So how can we detect this possible instability *a priori* ? It might seem natural to look at the determinant. But in the example above det(A) = 2, which is not extremely large nor small, so this is not the key. The magnification in the relative error is from two sources: x is in a direction with a large expansion under A and the error Δb is in a direction with a small expansion factor under A (and hence large expansion under A^{-1}). The maximum magnification factor, the condition number, is determined by the product of the largest expansion under A and the largest expansion under A^{-1}, i.e., cond(A) = $\|A\|\,\|A^{-1}\|$. Here $\|A\|$ = 100, $\|A^{-1}\|$ = 1/.02 = 50, so cond(A) = 5000.

Figure 9: *Solving Ax = b for an ill-conditioned matrix A*

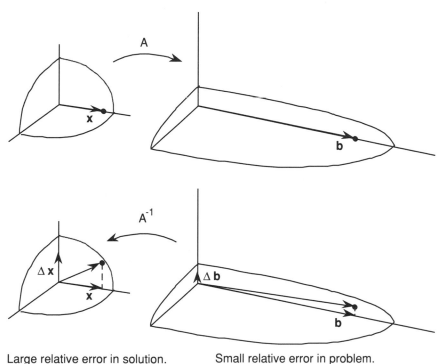

Large relative error in solution.   Small relative error in problem.

It is not the size (determinant) of the image ellipsoid that makes a matrix ill-conditioned, but rather its relative shape. Noting that $\|A\|\,\|A^{-1}\|$ = (largest σ_i)(1/smallest σ_i) points out once again the significance of the singular value decomposition and the geometry of Figure 7 in determining whether or not A is an ill-conditioned matrix. For the matrix of Figure 3, MATLAB yielded σ_1 =.6672, σ_2 = 2.2483, and cond(A) = 3.3699 (which is also the ratio of the lengths of the major and minor axes of the image ellipse). Hence this matrix is not ill-conditioned.

Least Squares Problems and Pseudoinverses

It is a common problem in statistics (regression) and numerical analysis to want to fit a "best" line (or curve) to a set of data points. Suppose that we wish to interpolate three points — (0,1), (1,0), (2,2) — with a straight line $y = mx + d$. Since we would really like each pair of coordinates to satisfy the equation, we need to solve the linear system

$$A x = \begin{bmatrix} m_0 + d \\ m_1 + d \\ m_2 + d \end{bmatrix} = \begin{bmatrix} 0 & 1 \\ 1 & 1 \\ 2 & 1 \end{bmatrix} \begin{bmatrix} m \\ d \end{bmatrix} = \begin{bmatrix} 1 \\ 0 \\ 2 \end{bmatrix} = b$$

This represents a linear transformation from $\Re^2$ to $\Re^3$. The image of the basic square (since x is in $\Re^2$) is a parallelogram which lies in the column space of A, a two-dimensional subspace of $\Re^3$. This is shown in Figure 10. If the vector b does not lie in this plane, then the system does not have a solution. A *least squares* solution s is a value of the vector x for which the sum of the squares of the y-distances, i.e., $\|b - Ax\|^2$, is a minimum; or equivalently, which minimizes the $\Re^3$ distance $\|b - Ax\|$. The shortest vector q in the set of possible values of s (there may be more than one, as in Figure 10) is referred to as the *pseudosolution* of $Ax = b$.

To determine this vector $\boldsymbol{q}$, we utilize the pseudoinverse A^+, which serves as a partial inverse for the full matrix A, and adapt the geometry of Figure 8 to Figure 10. A^+ maps $\Re^3$ onto $\Re^2$ in such a way that each $\boldsymbol{b}$ in $\Re^3$ maps to a vector $\boldsymbol{q}$ in $\Re^2$ with $A\boldsymbol{q} = \boldsymbol{p}$, the projection of $\boldsymbol{b}$ onto the column space of A. Then $\mathbf{q} = A^+\mathbf{b}$ is the pseudosolution of $A\mathbf{x} = \mathbf{b}$ since $\boldsymbol{p}$ is clearly the point of the column space with minimal distance to $\boldsymbol{b}$ and, in this case there is only one vector $\boldsymbol{q}$ with $A\boldsymbol{q} = \boldsymbol{p}$. The coordinates m_0 and d_0 of $\boldsymbol{q}$ yield the equation: $y = m_0 x + d_0$, which is the least squares line for the given data points. In this example $\boldsymbol{q} = (.5, .5)$ and the least squares line for the data is $y = .5x + .5$. We have shown the MATLAB computations for this data in Figure 15.

Once again, let us emphasize the general setting illustrated in Figure 8 and the related equations: $A\boldsymbol{s} = \boldsymbol{p}$ and $A^+\boldsymbol{t} = \boldsymbol{q}$. The problem and the solution form a symmetric pair. For this general case, $\boldsymbol{q}$ is the projection of each $\boldsymbol{s}$ onto the row space, and hence is the shortest vector with minimum $\|\boldsymbol{b} - A\boldsymbol{s}\|$, i.e. the least squares solution. (For a more direct comparison of the above example with the general case in Figure 8, the matrix Q_2^T can easily be expanded to operate on $\Re^3$ by adding the column $(0,0,1)^T$.)

Convergence of Regular Markov Chains

As a final application in which matrices play a central role, we look at finite Markov chains. In most situations purely algebraic methods are sufficient, but there is one case, the long-run equilibrium of what are called regular chains, in which geometric arguments similar to those given earlier lend insight.

For consistency with previous sections, we will use Strang's convention and let p_{ji} be the probability of moving to state j given that the current state is i. The column vectors of the transition matrix $P = [p_{ji}]$ are the probability distribution for the states (the row numbers) starting in the state associated with that column, and hence the columns add to one and the p_{ji} are nonnegative. (Usually the transpose is used, and the matrix is called *stochastic*.)

Figure 10: *Least squares, projection, and the pseudoinverse*

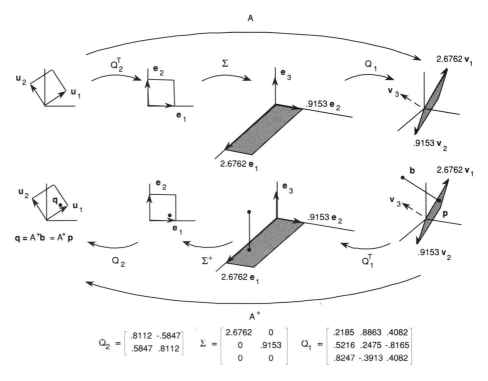

$$Q_2 = \begin{bmatrix} .8112 & -.5847 \\ .5847 & .8112 \end{bmatrix} \quad \Sigma = \begin{bmatrix} 2.6762 & 0 \\ 0 & .9153 \\ 0 & 0 \end{bmatrix} \quad Q_1 = \begin{bmatrix} .2185 & .8863 & .4082 \\ .5216 & .2475 & -.8165 \\ .8247 & -.3913 & .4082 \end{bmatrix}$$

Strang [13] gives a simple example of a Markov chain where there are two states: (1) a person lives outside of California, and (2) a person lives inside. Suppose that we know that every year 1/10 of the people outside move in, and 2/10 of those inside move out. The transition matrix for this chain is:

$$P = \begin{bmatrix} .9 & .2 \\ .1 & .8 \end{bmatrix}$$

We want to ask for the long run equilibrium, if any, in the two populations.

A Basic Tetrahedron for Transition Matrices of Markov Chains

When there are three states, the transition matrix P is a 3x3 matrix. The column vectors $c_i = (p_{1i}, p_{2i}, p_{3i})$ will lie in the first octant since $p_{ji} \geq 0$, so the image parallelepiped of the unit cube under the linear transformation $L(x) = Px$ is in the first octant. Moreover the endpoints of the column vectors always lie in the plane $x_1 + x_2 + x_3 = 1$, since the column elements add to one. The image parallelepiped thus always has vertices adjacent to **0** in this plane, just as the unit cube does. For this reason, instead of the unit cube we will use just the *basic tetrahedron* with **0** as vertex and the basis vectors e_1, e_2, and e_3 as edges, shown in Figure 11a. The image of this tetrahedron is the tetrahedron with the column vectors of P as its edges. One face of this image will always be in the plane $x_1 + x_2 + x_3 = 1$ and in the first octant, i.e., it is a subset of the corresponding face of the original basic tetrahedron. This is shown in Figure 11b for the matrix

$$P = \begin{bmatrix} .7 & 0 & .2 \\ .3 & .6 & .3 \\ 0 & .4 & .5 \end{bmatrix}$$

It is sufficient to look at the faces which are in the plane $x_1 + x_2 + x_3 = 1$. The triangular faces of Figure 11b are drawn as triangles in Figure 12a. We let the triangle formed by c_1, c_2, and c_3 represent the matrix P. This representation for a transition matrix was developed by Gabor Szekely [14] for studying factorization problems of stochastic matrices.

It is important in what follows to know when a vertex is not mapped strictly into the interior of the triangle. If column i has a zero in row j, then there is no j component of $c_i = Pe_i$, so the image of e_i lies on the edge opposite e_j. This happens in Figure 11 with c_1 and c_2. If the i th column of P has a one in the j th row (and so all the rest are zeros), then the image of vertex e_i is the vertex e_j. Consequently if the matrix has no zeros, then all the points of the image of the triangle will lie strictly within the interior of the triangle.

The matrix of probabilities of being in the various states after n time steps is P^n. In the California example the 2-step transition

■ Figure 11

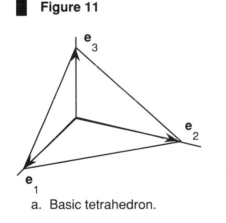

a. Basic tetrahedron.

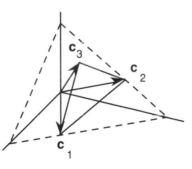

b. Image of the tetrahedron.

■ Figure 12

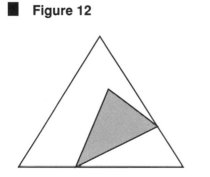

a. Image under P.

b. Image under P^2.

matrix is:

$$P^2 = \begin{bmatrix} .83 & .34 \\ .17 & .66 \end{bmatrix}$$

So 66% of the people in California will be there 2 years later (but some may have been away for one year).

If we iterate the transformation defined by a 3x3 transition matrix P, the image of each vertex of the triangle representing P is the same convex combination of the corresponding vertices of the previous triangle. For example

$$P^2 e_1 = p_{11} P e_1 + p_{21} P e_2 + p_{31} P e_3,$$

where $Pe_1 = p_{11} e_1 + p_{21} e_2 + p_{31} e_3$. The images of all the vertices will lie within the current triangle; and hence the image of the whole current triangle will lie within that triangle. But $P^2 e_1$ is also the first column vector of P^2. So this two-step transition matrix is represented by the triangle which is the image of the triangle representing P, and its vertices are the same convex combination of the vertices of the triangle representing P as before. The image of the original triangle under P^2 will be within the image under P. We see this in Figure 12b for our 3x3 example. Here the matrix P^2 is

$$\begin{bmatrix} .49 & .08 & .24 \\ .39 & .48 & .39 \\ .12 & .44 & .37 \end{bmatrix}.$$

Successive powers will give a sequence of nested triangles, each of whose vertices at every step is the *same* convex combination of the corresponding vertices of the previous triangle. The fact that this combination does not change with time is crucial to what follows.

Long Run Equilibrium of Regular Markov Chains

One of the fundamental results of finite Markov chains is that if a transition matrix P is *regular*, i.e., some power of P has all positive entries, then the powers P^n converge to a matrix all of whose columns are the same (hence rank one). This common column vector p_0 has positive entries which add to one, hence it is still a probability distribution. The vector p_0 can be interpreted as the long-run equilibrium probability distribution of the chain, since no matter what state the system starts in, the probability of ending up in each of the states after a large number of time steps converges to a fixed value.

The proof of this convergence is often omitted. There are proofs in Kemeny and Snell [8] and Roberts [12]. A key fact is that 1 is the dominant eigenvalue of P, and it has a single corresponding eigenvector which is positive. Since the column elements of P add to one, $\det(P^T - I) = 0$ where I is the identity matrix, so 1 is clearly an eigenvalue. But it is difficult to establish the rest. Groetsch and King [6] use Gershgorin's theorem to show that the dominant eigenvalue is one. The Perron-Frobenius theorem for nonnegative matrices (see Fiedler [3] and Luenberger [11]) says that the dominant eigenvalue has a positive eigenvector of multiplicity one. Once this is known, then the power method can be used to establish convergence of each column vector to p_0.

Our approach will be entirely geometric and visual. We are only trying to illustrate the plausibility of the convergence to p_0. This result is not self-evident.

First, consider the two state case. Here the line segment joining e_1 and e_2 replaces the triangle of Figure 12. The images $c_1 = Pe_1$ and $c_2 = Pe_2$ are on

Figure 13

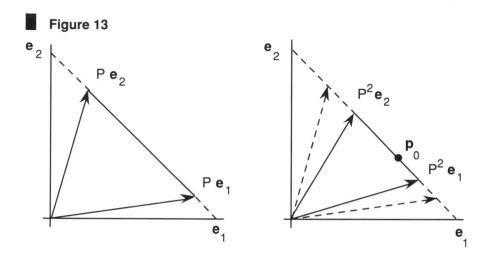

this line segment, and every point on the line segment is mapped into the line segment joining Pe_1 and Pe_2. The 2x2 California example is drawn in Figure 13.

Here both e_1 and e_2 are mapped into the interior of the line segment, so the line shrinks, and successive powers shrink by the same factor. Since the image of e_1 moves in .1 of the remaining distance, and e_2 by .2, it is also plausible that they should meet in the ratio 1:2 — i.e., at (2/3,1/3). Strang shows, by finding the eigenvalues (1 and .7), that in fact P^n converges to the matrix with column vectors (2/3,1/3).

Now back to the 3x3 case. The key idea is that the image triangles will shrink to a point. This limit point is the equilibrium distribution. When can this occur? It is not enough for the area (or volume) to go to zero; the lengths of the sides must also go to zero.

In the case of a regular chain, for some integer m, the images of the vertices will break away from the edges of the triangle, since there will be no zeros in P^m. Now consider powers of $Q = P^m$. Since the triangles are nested, if powers of Q converge to a point, so will powers of P. As seen before, Q maps the triangle strictly into its interior, and successive powers of Q are strictly inside of the previous image by the same factors. We next show that the triangles are forced to shrink to a point.

To see this, let $\varepsilon > 0$ be the smallest entry of Q. Recall that each new vertex is the same convex combination of the old vertices at each stage. Suppose vertex one is mapped to $(1 - \alpha - \beta)v_1 + \alpha v_2 + \beta v_3 = v_1 + \alpha(v_2 - v_1) + \beta(v_3 - v_1)$, where α, β, and $1 - \alpha - \beta$ are at least 0. It then follows for the matrix $Q = P^n$ above that each vertex of the new triangle must lie within the dotted triangle in Figure 14, since α, β and $1 - \alpha - \beta$ are all at least ε. (Because we can apply this argument to any two vertices, $1 - \alpha - \beta \geq \varepsilon$ means that the vertex cannot get beyond the opposite dotted line.) This forces the longest side of the new triangle to be at most $(1 - 2\varepsilon)$ times as long as the longest side of the previous triangle. We have $0 \leq (1 - 2\varepsilon) < 1$ because $0 < \varepsilon \leq .5$. Since this happens at each stage of the sequence, the sides of the nested triangles must converge to 0, and thus the closed, nested triangles must converge to a point p_0 strictly in the interior of the original triangle. This means that all of the columns of the limit of $Q^k = P^{km}$, and so of P^n (since the sequence is nested), will converge to the positive vector p_0.

This limit vector p_0 turns out to be the fixed point of the mapping defined by P, and hence the single eigenvector for the eigenvalue one. We can see this from the geometry, since its image under P must be in each of the triangles of the sequence, so it can only be p_0 itself. This is one way to compute p_0: solve $Px = x$, with $x_1 + x_2 + x_3 = 1$. There is exactly one solution. For the 3x3 example we get $p_0 = (.2286, .4286, .3430)$. Another, less efficient, way is to compute successive images of some probability vector under P until the result

Figure 14

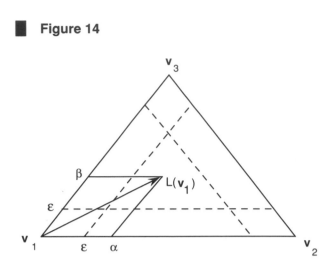

stabilizes. This could be done with a system such as MATLAB, for example. This amounts to the power method for finding the dominant eigenvector.

It's not hard to visualize the four-state case, and to consequently believe the general result. The triangle will be replaced by a tetrahedron, and we will still have that each new image under Q is a tetrahedron which lies entirely within the tetrahedron formed by shaving off a slab from each side of the previous tetrahedron so that the fraction ε is cut off the end of each edge.

Computations with MATLAB

Most of the computations were done with MATLAB [15] (which stands for MATrix LABoratory), an interactive scientific computation system based on the state-of-the-art LINPACK and EISPACK libraries. It is recognized as the premier interactive program for numerical linear algebra and matrix computation. It is available on a wide range of machines, from a Macintosh or PC up to a Cray.

It is accurate, fast, and exceedingly easy to use. In Figure 15 we present a complete session, with comments, of MATLAB on a Macintosh SE. This figure contains computations for the data of Figure 10. MATLAB is not only useful for practicing scientists and engineers, but for instructional use as well. We have used it in undergraduate courses and found that the students had no trouble adjusting to it. The computations of this paper are usually much too tedious to do seriously by hand in a class. With this kind of power available, we wonder how we will change the way we teach linear algebra, and what topics we will choose.

Summary

We firmly believe that geometry is a powerful ally in the study and understanding of the ideas of linear algebra. We have been strongly influenced by the book of Strang [13], which we use in a second course. His use of geometry for both motivation and for gluing together various ideas has been a strong factor in our using visualization as a tool for gaining understanding.

We hope that you will find the the basic unit cube, unit ball, eigencube, and tetrahedron to be useful objects for illustrating the effects of a linear transformation, and that this will encourage the use of geometry in the study and teaching of linear algebra. In addition, we hope that you have gained an appreciation for the central role of the matrix $A^{\mathsf{T}}A$ and the singular value decomposition, $A = Q_1 \Sigma Q_2^{\mathsf{T}}$, in the study of linear transformations. Finally, we have illustrated, for a difficult problem in Markov chains, the insight gained through looking at the geometry in a new way.

Figure 15: *MATLAB session*

```
»A = [0 1                    Enter  the data after the MATLAB prompt:  »
1 1
2 1]                         Note how the matrix A can be easily entered in a natural way.
A =
    0   1                    MATLAB responds.
    1   1
    2   1
»b = [1 0 2]'                Enter the column vector b.
b =
    1
    0
    2                        =====

»q = A\b                     Solve Ax = b.
q =                          The least squares solution q is returned, since there is no exact solution.
    0.5000
    0.5000

»p = A*q                     The projection p of b onto the column space of A.
p =
    0.5000
    1.0000
    1.5000                   =====

»[Q1, sigma, Q2] = svd(A)    Find the SVD of A: Q₂ᵀ sigma Q₁
Q1 =
    0.2185   0.8863   0.4082
    0.5216   0.2475  -0.8165
    0.8247  -0.3913   0.4082
sigma =
    2.6762      0
       0   0.9153
       0      0
Q2 =
    0.8112  -0.5847
    0.5847   0.8112            =====

»pinv(A)                     Find the pseudoinverse of A.
ans =
   -0.5000  -0.0000   0.5000
    0.8333   0.3333  -0.1667

»pinv(A)*b                   Solve Ax = b using the pseudoinverse.
ans =                        This agrees with q above.
    0.5000
    0.5000                     =====

»A'*A                        Find the singular values as square roots of the eigenvalues of AᵀA.
ans =
    5   3
    3   3
»eig(A'*A)
ans =
    7.1623
    0.8377
»sing = sqrt(ans)            This is element-wise square root.  Note that you can use the variable ans.
sing =
    2.6762
    0.9153
```

References

[1] Anton, Howard and Chris Rorres, *Elementary Linear Algebra with Applications*, John Wiley and Sons, 1987.

[2] Blank, S. J., Nishan Krikorian, and David Spring, "A Geometrically Inspired Proof of the Singular Value Decomposition," *Amer. Math. Monthly*, 96(1989), 238-239.

[3] Fiedler, Miroslav, *Special Matrices and Their Applications in Numerical Mathematics*, Martinus Nijhoff Publishers, Netherlands, 1986.

[4]   Forsythe, G. E., M. A. Malcolm, and C. B. Moler, *Computer Methods for Mathematical Computations*, Prentice-Hall, 1977.

[5]   Green, Paul E., *Mathematical Tools for Applied Multivariate Analysis*, Academic Press, 1976

[6]   Groetsch, C. W. and J. Thomas King, *Matrix Methods and Applications*, Prentice-Hall, 1988.

[7]   Hoechsmann, K., "Singular Values and the Spectral Theorem," *Amer. Math. Monthly,* 97(1990), 413-414.

[8]   Kemeny, John G. and J. Laurie Snell, *Finite Markov Chains*, Van Nostrand, 1960, reprinted by Springer-Verlag, 1976.

[9]   Long, Cliff, "Geometry of Some Linear Transformations." A super-8 film produced at Carleton College under an NSF-funded workshop, "Computer Graphics for Learning Mathematics," administered by Roger Kirchner, 1975.

[10] Long, Cliff, "Visualization of Matrix Singular Value Decomposition," *Math. Magazine,* 56(1983), 161-167.

[11] Luenberger, David G., *Introduction to Dynamical Systems*, John Wiley and Sons, 1979.

[12] Roberts, Fred S., *Discrete Mathematical Models with Applications to Social, Biological, and Environmental Models*, Prentice-Hall, 1976.

[13] Strang, Gilbert, *Linear Algebra and Its Applications*, Third Edition, Harcourt Brace Jovanovich, 1988.

[14] Szekely, Gabor, University of Eötvös, Hungary. Private communication.

[15] *MATLAB*™, software by The MathWorks, Inc., Natick, MA.

The Vector Field Approach in Complex Analysis

Bart Braden

Picturing a complex function $f(z)$ as a vector field on the complex plane gives a clear geometric interpretation of complex integrals comparable to the interpretation of a real definite integral as the area under the graph of the function over an interval. Moreover, the vector field associated with $f(z)$ provides a good picture of important features of the function f, such as the location and order of its zeros and poles, the location of any branch points or essential singularities and the behavior near these points. These ideas, originally due to G. Pólya [3], are discussed in [1] and [2].

The usefulness of the vector field interpretation in analyzing a specific function or integral depends on the amount of work required to draw the appropriate vector field. Recently developed graphics workstations and software make it rather easy to write interactive programs to display the vector field for any computable function, and such programs can be useful in teaching and learning complex analysis. This article summarizes the results detailed in [1] and [2], and briefly describes some more recent developments. The graphics images were produced using the interactive graphics/computation program $Mathematica^{TM}$.

Since a complex function f can be thought of as a function from $\Re^2$ to $\Re^2$, it is natural to interpret f as a vector field on the plane: at each point z attach the plane vector $f(z)$. This idea turns out to be of limited utility. Polya's insight was that instead one should attach the conjugate $\overline{f(z)}$ at each point z. Thus to any complex function f we associate the vector field $\overline{W}(z) = \overline{f(z)}$. The conjugation plays a key role, making the analytic properties of f correspond to simple geometric properties of $\overline{W}$. Note that the length of $\overline{W}$ tells us the modulus $|f(z)|$ and the polar angle of $\overline{W}(z)$ is the negative of the argument of $f(z)$. Actually, because the variation in modulus is typically several orders of magnitude, our plots have the lengths scaled logarithmically. The following results are shown in [1]:

i) The function f is analytic in a region exactly when $\overline{W}$ is differentiable, divergence-free and curl-free throughout the region. (This is a geometric interpretation of the *Cauchy-Riemann equations*.)

ii) If z_0 is a zero of f of order k, then the index of $\overline{W}$ at z_0 is $-k$. That is, as z traverses a small circle around z_0 the vectors $\overline{W}(z)$ turn *clockwise* through k revolutions. Similarly if z_0 is a pole of f of order k then the index of $\overline{W}$ at z_0 is k: the vectors $\overline{W}(z)$ turn *counterclockwise* through k revolutions as z circles z_0. Finally, if f is analytic and nonzero at z_0, then the index of $\overline{W}$ at z_0 is zero. (This is a geometric interpretation of the *principle of the argument*.)

Hence from a plot of $\overline{W}$ in a region we can spot the zeros and poles of f in that region, and can visually determine the order of each zero or pole. For

Bart Braden is Professor of Mathematics at Northern Kentucky University. He received his Bachelor's degree from Washington State University followed by graduate degrees from Yale and the University of Oregon. His interests include classical applied mathematics, symbolic computation and computer graphics, and Chinese language and culture.

Figure 1: *The Pólya vector*

field of $f(z) = \dfrac{(z-1)^2}{z^2+1}$

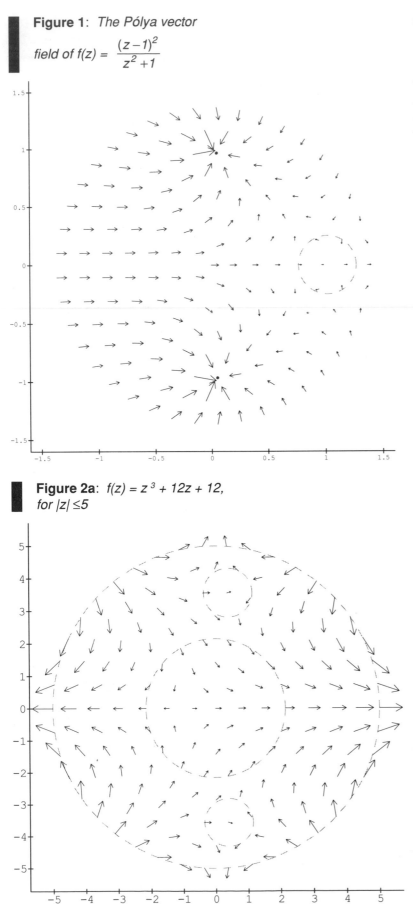

Figure 2a: $f(z) = z^3 + 12z + 12$,
for $|z| \leq 5$

example, consider the rational function

$$f(z) = \frac{(z-1)^2}{z^2+1},$$

which we know has simple poles at $\pm i$ and a zero of order 2 at $z = 1$. This can clearly be seen in the corresponding vector field $\overline{\mathbf{W}}$ in Figure 1. Note that the arrows are only slightly shorter near the zero, a result of our logarithmic scaling, but the nonzero index of the vector field at this singular point is apparent. The long arrows near the poles make them stand out.

An especially helpful feature of *Mathematica* here is that the graphic images can be resized simply by dragging a sizing bar with the mouse. Thus any portion of the screen image can be magnified as much as desired, which unfortunately is not the case with our printed image.

An interactive program to draw the vector field makes an excellent tool for locating the zeros of polynomials or other complex functions. Using the principle of the argument one first finds a disk containing one or more zeros, and then successively zooms in on smaller and smaller disks containing a zero. Figures 2a and 2b show this process applied to locate one of the complex zeros of the cubic $z^3 + 12z + 12$ using our program ZOOM, which draws the vector field $\overline{\mathbf{W}}$ in the disk with any user-specified center and radius. In Figure 2a the real zero and two conjugate complex zeros are apparent as singular points where the vector field has index -1. Note that the vectors on the outermost ring turn three revolutions clockwise as we traverse this ring counterclockwise which means, by the principle of the argument, that f has three zeros somewhere in this ring.

In Figure 2b the disk of radius .2 centered at $.5 + 3.5i$ is analyzed, and one sees the behavior near the zero in this disk more clearly. *Mathematica* allows the user to read the coordinates of the cursor, so by moving the cursor to make it appear to lie over the singular point of the vector field in Figure 2b, we estimate that the complex zero of our polynomial in the upper half-plane is approximately $.47 + 3.56i$. If greater accuracy were desired we could use this

Figure 2b: $f(z) = z^3 + 12z + 12$, for $|z - (.5+3.5i)| \leq 0.2$

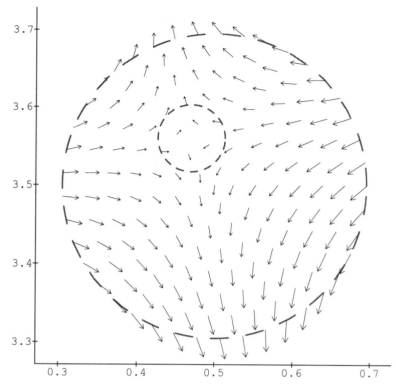

Figure 3a: $f(z) = Sqrt[z-1] * Sqrt[z+1]$

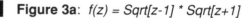

value as a new center, and run the ZOOM program once more with a radius of say .05. Alternatively, we could use it as the starting point for the *Mathematica* FindRoot command, i.e., Newton's method.

Not only does this interactive graphical search make the complex zeros seem more "real" to the student; he or she also soon notices that at high magnification zeros of analytic functions of a specified order k all look essentially alike, regardless of the specific function involved. Analysis soon reveals the reason. $f(z)$ has a zero of order k at z_0 exactly when the leading term in its Taylor expansion about z_0 is of degree k: $f(z) = c(z - z_0)^k + O(z - z_0)^{k+1}$, where c is nonzero. Thus the Pólya vector field of f looks like that of $c(z - z_0)^k$ very near z_0. If $\overline{\mathbf{V}}$ is the Pólya vector field of $(z - z_0)^k$, then since multiplication by $\overline{c}$ simply magnifies the length of $\overline{\mathbf{V}}(z)$ by $|c|$ and turns it through the angle $-\text{Arg}(c)$, we see that $\overline{\mathbf{W}}$ differs from $\overline{\mathbf{V}}$ near z_0 only by this magnification and rotation.

The same reasoning applies to the behavior near a pole of order k: the vector field $\overline{\mathbf{W}}$ of f differs from that of $(z - z_0)^{-k}$ only by a change of scale and a rotation. In particular this applies to simple poles of f, where the leading term in the Laurent expansion of f is $c/(z-z_0)$, c being the residue of f at z_0. Since the Pólya vector field of $1/(z-z_0)$ is radially outward with center z_0 (see [2]), we have a simple geometric interpretation of the residue of f at a simple pole z_0: the vector field $\overline{\mathbf{W}}$ of f near z_0 looks like the radial vector field of $1/(z-z_0)$, only magnified by the factor $|\text{Res}(f,z_0)|$ and turned through the angle $-\text{Arg}[\text{Res}(f,z_0)]$. In Figure 1, for example, the radially inward vectors near the simple poles at i and $-i$ tell us immediately that the residue of f at each of these poles is a negative real number. (The logarithmic scaling of the lengths of the arrows in this plot does not allow us to estimate the modulus of the residue accurately at these poles.) The residue of f at a pole of higher order also can be estimated geometrically from a plot of $\overline{\mathbf{W}}$ near the pole, as we explain below, but the estimation requires more care.

The concepts of branch points and branch cuts for multivalued complex functions are

■ **Figure 3b**: *f(z) = Sqrt[z²-1]*

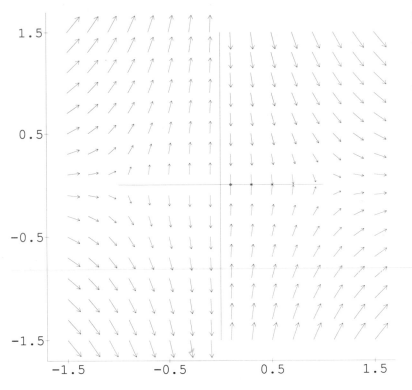

■ **Figure 4**: *The Pólya vector field of f(z) = z² along the unit circle*

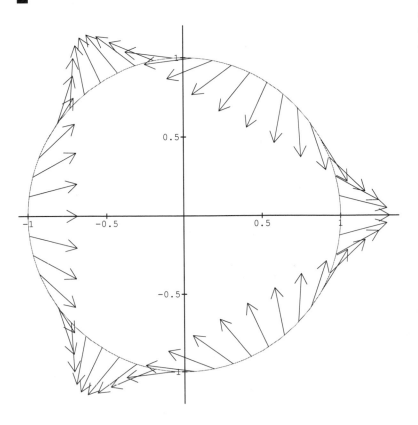

notoriously difficult for beginning students in complex analysis. An interactive program to draw the Pólya vector field can be very helpful in clarifying these ideas. The vector field $\overline{W}$ of a single-valued branch of a multivalued function clearly displays branch lines and the behavior near branch points. This is seen in Figures 3a and 3b, where two branches of the square root of $z^2 - 1$ are shown. The discontinuity of the vector fields along the branch lines is apparent.

Requiring students to define a branch of $f(z)$ so that the program ZOOM will produce a plot of a vector field with specified branch lines gives them a clear geometric goal. When the task of defining branches is phrased this way, it is perhaps easier for beginning students to comprehend the problem and to know when they have achieved success.

Perhaps the most significant use of the vector field picture of complex functions is in visualizing contour integrals. This is described in detail in [2], so we just summarize the results here. To estimate the value of

$$\int_\gamma f(z)dz$$

visually, one begins with a plot of the associated vector field $\overline{W}$ along the contour γ as in Figure 4. The real part of the integral turns out to be simply the integral of the tangential component of $\overline{W}$ along γ and its imaginary part is the integral of the normal component of $\overline{W}$ along γ.

In Figure 4 we see from the symmetry around the real axis that the tangential components of $\overline{W}$ at $e^{i\theta}$ and $e^{-i\theta}$ are negatives of each other, so the real part of the integral of $f(z) = z^2$ over the circle is zero. Similarly because the vectors on opposite ends of any diameter of the circle are equal (so their normal components are negatives of each other), the normal component also integrates to zero. It follows, then, that

$$\int_\gamma z^2 dz = 0 \ .$$

The analytic proof of this fundamental result is quite simple, and has the advantage that it is easily generalized to show that

$$\int_\gamma z^n dz = 0$$

for any $n \neq -1$. However, I believe the visual proof still has pedagogical value: the argument from symmetry may be more memorable than a computational proof. (Compare this with the argument in calculus that the integral of an odd function over an interval $[-a,a]$ is zero.)

Just as the interpretation of definite integrals of real functions as areas is useful as a check on analytical calculations, so the above geometric interpretation of complex integrals can be used to catch blunders when the integrals are calculated via the residue calculus. (See [2] for examples.) In any case the concrete geometrical interpretation dispels the aura of unreality which often surrounds complex contour integrals in students' minds.

Since the residue of f at an isolated singular point z_0 is the value of the integral

$$\frac{1}{2\pi i} \int_\gamma f(z) dz \,,$$

where γ is a small circle around z_0, we can evaluate residues geometrically by visually estimating the integral of f around a small circle $|z - z_0| = \varepsilon$. Of course the primary use of the residue is for evaluating this integral, so our argument is circular. But the point is, whereas the residue at a simple pole can easily be estimated directly from the vector field (as discussed above), at poles of higher order or at essential singular points the residue can be estimated only by visually integrating the tangential and normal components around the singular point. (Lower order terms in the Laurent series mask the term involving the residue, but they all integrate to zero.)

Besides its pedagogical uses, the vector field picture of complex integrals can lead to new theoretical insights. For example, in [2] a simple necessary and sufficient condition is derived for equality to hold in the fundamental triangle inequality for integrals:

$$\left| \int_\gamma f(z) dz \right| \leq \int_\gamma |f(z)| ds$$

namely, the angle between the vector field $\overline{\mathbf{W}}$ and the tangent vector to γ must remain constant along the curve. This geometric condition is not conveniently expressible in terms of the mapping picture of the function f. Perhaps other theoretical questions about complex integrals will benefit from this sort of geometric analysis.

A final application, suggested to me by Alan Gluchoff, is to use the vector field picture of complex functions to clarify various ideas involving convergence of sequences of complex functions. Here the ability of *Mathematica* to animate a sequence of graphics images can be brought into play with striking effect. For example, the fact that a power series converges inside its circle of convergence and diverges outside can be demonstrated in specific cases by plotting the Pólya vector fields for the partial sums of the series and then animating the resulting sequence of images. Inside the circle of convergence one sees that the partial sums quickly stabilize, while outside this circle the vectors representing the partial sums continue to spin and grow in length, indicating divergence of the series. Moreover, one observes that the rate of convergence or divergence is very rapid far from the circle of convergence, but it slows as one approaches this critical circle. Students who have seen these animated displays say the visual experience makes the existence of the circle of convergence of a power series more vivid in their minds than the formal proof of its existence or even its calculation in specific instances by the ratio or root test. Providing such direct perceptions of mathematical truths is the aim of visualization efforts in mathematics education. The vector field approach, coupled with modern computer hardware and a graphics/computation system such as *Mathematica*, now brings this goal within our grasp in many parts of classical complex analysis.

Note on Software

The graphics images in this paper were produced using interactive *Mathematica* notebooks written by my student Larry Menzer and me for a Macintosh II computer with 4 or more megabytes of RAM. They are available upon request. *Mathematica*™ (Version 1.2) is available from Wolfram Research Inc., P.O. Box 6059, Champaign, IL 61821.

References

[1] Braden, Bart, "Picturing Functions of a Complex Variable," *The College Mathematics Journal*, 16(1985) 63-72.

[2] Braden, Bart, "Pólya's Geometric Picture of Complex Contour Integrals," *Mathematics Magazine*, 60(1987) 321-327.

[3] Pólya, George, and Gordon Latta, *Complex Variables*, Wiley, 1974.

Using Fractal Images in the Visualization of Iterative Techniques From Numerical Analysis

Valerie A. Miller and G. Scott Owen

The use of graphical images has been traditional in teaching lower level mathematics. When teaching basic arithmetic, images of physical objects such as apples or pies are often used to give students a visual reinforcement of the quantities being manipulated. When a student is learning to apply a mathematical principle to a word problem in an algebra or calculus class, the teacher instructs the student to "draw a picture" of the problem, if possible. This suggestion is an effort to induce the student to visualize the problem in concrete terms and thus envision possible avenues for solving it.

However, when a student enters a mathematics class beyond the calculus level the use of visual reinforcement of mathematical concepts quickly becomes nonexistent, except for the occasional graph theorist's graph (composed of vertices and edges) or the plotting of a function. This lack of visualization is not all that surprising when you think of trying to "draw" an abelian group, a Banach algebra, or a differential equation. Until recently, before the formal development of fractal geometry [4], the idea of visualizing a mathematical iterative process (in other than tabular form) was also inconceivable. The well known Mandelbrot and Julia sets are both generated with the iterative scheme

$$z_{k+1} = z_k^2 + c$$

(under suitable initial and terminating conditions) and produce extraordinary pictures, but neither seems to be useful in relaying to the student how or why a numerical mathematician selects one specific iterative scheme over another to solve a particular problem. The two main areas of consideration when selecting an iterative method are the robustness and the complexity of the algorithm. By robustness of an algorithm we are referring to the types of problems on which the algorithm will work and for what intervals the algorithm will converge. By complexity of an algorithm we mean the speed of convergence and the cost per iteration. In this paper we present ways in which the natural tie between fractal image generation and iterative methods can be used to improve the student's understanding of these theoretical considerations.

There are several different, yet common, situations where iteration is used as a means of solution, e.g., solving $f(x) = 0$, solving linear systems of equations (both via splittings and iterative refinement), eigenvalue approximations, solving nonlinear systems of equations, as well as adaptive quadrature methods. Thus, there are many types of iterative processes that we could choose from to illustrate our techniques. For simplicity, we consider the most basic of mathematical problems: determine a complex number x such that $f(x) = 0$.

Valerie Miller is Assistant Professor of Mathematics and Computer Science at Georgia State University. Her research interests include numerical analysis, linear algebra, parallel processing, and visualization.

G. Scott Owen is Professor of Mathematics and Computer Science at Georgia State University. His research interests include computer graphics, visualization, computer science education, computer graphics in education, software engineering, and artificial intelligence.

When attempting to solve $f(x) = 0$, the function f need not be very complicated before simple analytical methods become inadequate. For example, there are formulae for solving polynomial equations of degree less than or equal to four, but none such exist for higher degrees. Thus, unless the polynomial function factors into a product of polynomials whose degrees are four or less, some other procedure, such as iteration (usually done on a computer), is necessary for solving the problem. Any student of numerical analysis knows there are many different iterative schemes for solving $f(x) = 0$. Each of these schemes has an order of convergence associated with it (a number that indicates how quickly, in terms of iteration count, the scheme will converge to a solution) as well as the requisite hypothesis for when it will converge to a solution.

A student is exposed to many such iterative schemes and has the problem of deciding which is best. The student might initially opt for the "quickest" method, defined as the scheme that converges to a solution in the fewest number of iterations. This may not always be the best choice. High order methods usually require derivatives and these may not be available. If the necessary derivatives are available then the computational cost of a single iteration of a high order iterative scheme may cause it to be more computationally expensive than another scheme that requires more, but cheaper iterations. Thus, the cost per iteration as well as the count of iterations must be considered when selecting an iterative technique. With regards to the robustness of an algorithm, the theoretical considerations for the determination of the interval of convergence may be difficult to ascertain and/or verify for many problems. It is also the case that even if the theoretical interval of convergence is known it can be quite narrow due to the various inequality arguments necessary for its derivation. On the other hand, the numerical interval of convergence can be much larger than its theoretical counterpart. Hence, the ability to visualize some type of performance profile for an iterative process would help the student compare techniques in terms of their rate of convergence, cost per iteration and interval of convergence and decide which method is more appropriate for a particular type of problem.

As there are many iterative schemes available to solve $f(x) = 0$ we have selected four to use in our illustrations: the secant method (rate of convergence ~ 1.62), Newton's method (rate of convergence = 2), Halley's method (rate of convergence = 3), and Aitken's acceleration technique applied to the secant method (rate of convergence ~ 2.24). These methods were selected on the basis of their frequent usage and various rates of convergence. Detailed derivations and theoretical considerations for these algorithms can be found in any numerical analysis text (e.g., [1], [3]) and so will not be discussed here.

Since iterative techniques simply generate a sequence of approximations to a solution of $f(x) = 0$, a stopping criterion is needed to terminate the method when an acceptable answer has been found. The two criteria that we will use are:

$$1) \qquad |f(x_k)| < \text{tolerance} \qquad\qquad (1.1)$$
$$2) \qquad |\,|x_{k+1}|^2 - |x_k|^2\,| < \text{tolerance} \qquad (1.2)$$

Criterion (1.2) will be used most often in our discussions as it has been used previously in this type of presentation ([2]) and it generates more interesting images (though qualitatively, the two criteria generate images that contain essentially the same information). The value of "tolerance" that one uses depends on the accuracy desired as well as the precision available on the computer or calculator that is being used to perform the computation. In our examples we use .0001 as the value of tolerance.

Before proceeding with the discussion on the generation and usage of our images, we will point out how we are going to use Aitken's acceleration technique as this can be implemented in a variety of ways. This method uses three successive iterates x_i, x_{i+1}, and x_{i+2} of a given method F and algebraically manipulates them by the formula

$$a_i = x_i - \frac{(x_{i+1} - x_i)^2}{x_{i+2} - 2x_{i+1} + x_i}$$

into a (usually) better approximation than that which method F would have generated. One way this Aitken iterate can be used is to embed it into the sequence generated by method F; i.e., if x_1, x_2, and x_3 are the three iterates from method F and a_1 is the Aitken iterate generated by these, then one sets $x_4 = a_1$ and proceeds to generate the next iterate x_5 by method F.

As we will use Aitken's method to accelerate the convergence of the iterates from the secant method, which requires two iterates at each step in order to generate a third, this type of acceleration is not really appropriate. So we will abide by the theory behind acceleration techniques and generate a separate sequence of Aitken iterates based on the given sequence of secant iterates. It should be noted that generating the Aitken iterates in this fashion will increase the cost of the Aitken iteration in our examples. Aitken's acceleration technique does not have a fixed order of convergence since it depends on the method being accelerated. It can be shown that if F is an iterative method of order $p > 1$ then Aitken's acceleration technique applied to F is a method of order $2p - 1$ [3].

In the remainder of the paper we will concentrate our discussion on the following three topics:
1) how one can create and use images that will aid students in visualizing how iterative schemes work;
2) what it means to have different rates of convergence; and
3) that in terms of cost, it is not always best to use a high order of convergence method to solve a problem.

Creating the Visualization Aids

As illustrated in Pickover's paper [2], an iterative method for solving $f(x) = 0$ for x in the complex plane **C** can be represented graphically in the following manner. First, select a square region in **C** where you think the roots of $f(x)$ lie. This region determines the window coordinates for your graphics device. Each pixel can be made to correspond to a unique world coordinate value from the selected region of **C**, by using the natural discretization mapping from this region to the graphics device. We select as our initial guess of a root of $f(x)$ the world coordinate associated with a pixel. With this as our x_0 we then proceed to perform the iterative process until either we generate an iterate, x_k, that satisfies our stopping criterion or until a maximum number of iterations has been exceeded. Either way we use this value of k to "color code" the pixel that is associated with x_0. These color-coded pixels present a fractal image representing the iterative method's results for solving $f(x) = 0$. We call these graphical representations fractal images since they evince self-similarity, if explored to sufficient detail [2], have regions of chaotic behavior, and have

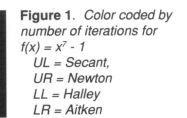

Figure 1. *Color coded by number of iterations for f(x) = x⁷ - 1*
  UL = Secant,
  UR = Newton
  LL = Halley
  LR = Aitken

previously been given as examples of fractal images [5]. We present two color plates in which each contains an original image and three enlarged regions so that the self-similarity might be made more evident.

This color coding allows the student to observe how quickly, in iteration count, an algorithm converges to a root of $f(x)$ from various starting points. The student can generate images for each method as described above (using identical stopping criteria and color maps) and then compare the different methods by examining the various colorizations of a particular starting iterate. It is even more informative to generate a single screen in which several methods are presented so that the student can look at the single screen and perform a simultaneous visual comparison of the different methods.

Using the Images

In Figure 1 we give an example of this type of image. Four methods are used to solve the problem $x^7 - 1 = 0$ in the region [-5.0, 5.0] x [-5.0, 5.0]. The method being represented in the upper left is the secant method; the upper right is Newton's method; the lower left is Halley's method; and the lower right is Aitken's method. The stopping criteria used in all of these methods was (1.2) with tolerance equal to 10^{-4}. It is the use of this particular criterion that is responsible for the "spiny" look of each of the petals. It is also this stopping criterion that is responsible for the "ribbons" found in the corners of the Aitken's method fractal. These ribbons can be interpreted as being places where the method yields a type of "false" convergence. By this we mean that

Figure 2. *Color coded by cost per iteration for*
$f(x) = x^7 - 1$
  UL = Secant
  UR = Newton
  LL = Halley
  LR = Aitken

the difference in the norms of two successive Aitken's iterates has become small very quickly, but the iterates have probably not truly converged to a root of the polynomial (the image should contain a certain amount of continuity). If a more standard stopping criterion, e.g., (1.1), is used the petals become much more rounded and the aforementioned ribbons disappear (Figure 3).

We have used two different color schemes for our images. While we generate and display the images to the students in color, the figures located in this paper use only a gray scale. Since there are about 12 shades of gray that can be easily distinguished by the naked eye, this is all we use. As we also have iteration counts greater than 12, the coloring of the pixels is done with mod 12 arithmetic. In general this does not present a problem as iteration counts increase smoothly as the starting iterate gets further from the origin. Where this could present a problem is in the examination of the rays containing the seventh roots of -1, which is where the self-similarity occurs. The gray scale is from dark to light; i.e., the darker shades are for small iteration counts (mod 12) and the lighter shades are for larger iteration counts (mod 12).

For Color Plates 4 and 5, we used the color table listed below with the color number corresponding to the number of iterations:

color number	color(s)
1-10	shades of red (1 = dark red, 10 = light red)
11-20	shades of green (11 = dark green, 20 = light green)
21-30	shades of blue (21 = dark blue, 30 = light blue)
31-40	fixed green mixed with the 10 shades of red
41-50	fixed blue mixed with the 10 shades of green
etc.	

Table 1: The Color Table.

The use of shades of primary colors and the knowledge of the order in which they occur in the color table makes it easy to see where a method converges to a solution quickly (the first time through the gray scale in the black and white figures or the red hues in the color plates) and where it converges more slowly (the second or third repetition of the gray scale in the black and white figures or the blue hues in the color plates).

Figure 3. *Color coded by number of iterations for*
$f(x) = x^7 - 1$
  Criteria, $|f(x^k)| < tolerance$
  UL = Secant, •
  UR = Newton
  LL = Halley
  LR = Aitken

In Figure 1 the highest ordered method, Halley's method (in the lower left), has only the first set of gray shades in the majority of its representation, indicating that it converges to a solution in 12 or fewer iterations for most starting iterates in our specified region. Newton's method (upper right), one of the most commonly used iterative techniques, can be seen to converge more slowly than Halley's method since its image contains a second set of gray shades. This indicates that for many points in the region Newton's method requires more than 12 iterations to obtain convergence. As Newton's method is only a quadratically convergent scheme this is to be expected. Similar comments can be made for the remaining methods. Thus, by using this type of visualization, it is now clearer to the student what it means to say that "the higher the order of convergence of an iterative scheme, the fewer iterations necessary for convergence to a solution."

Our first example of the self-similarity in these images is Color Plate 7. In this image the upper left-hand corner is the colorized version of Newton's method for solving $x^7 - 1 = 0$ (which was the upper right-hand corner in Figure 1) on [-5,5] x [-5,5]. The other three images are representations of the regions:
upper right
  [-5,-3.7:5] x [-0.625,0.625],
lower left
  [-5,-4.7] x [-0.15,0.15], and
lower right:
  [-4.8,-4.7] x [0.1,0.2] .

Note the recurring chains of "nodules" (as Pickover refers to them in [2]). The colorization of these nodules changes as the region becomes smaller as the number of required iterations necessary for convergence increases. However, it is the repeated pattern that is of more interest here as it is in this pattern that the self-similarity is contained. Notice also that at all scales there are regions of order and regions of chaos.

When considering a high order iterative method for solving a particular problem there is usually a penalty, in terms of the cost per iteration, for its rapid convergence to a solution. The cost of an algorithm is subjective, though a commonly accepted means of evaluating this is in terms of the amount of CPU time required to perform a particular operation. As this cost is machine and language compiler dependent, a numerical analyst often uses the simple (and easily understood) algorithm analysis method of estimating the cost of an iterative process by the number of floating point operations (flops) required per iterate. If a more thorough study of the cost of an algorithm is desired (as would be appropriate in an analysis of algorithms class) then a credit analysis of the algorithm can be performed. In a credit analysis the cost of the worst case scenarios together with the cost of good case scenarios and time necessary for execution are considered (for a discussion on credit analysis and amortized cost see [6]). In our examples of cost comparisons between the four methods we used the simple analysis method and the following table of estimated costs:

Operation	Cost (in flops)	
Addition	1	
Subtraction	1	
Multiplication	3	
Division	6	
x^k	$3(k\text{-}1)$	(k an integer)
$\sin(x)$	25	
$\exp(x)$	50	

Table 2: The Costs of Arithmetic Operations

When solving $f(x) = 0$ the higher ordered iterative methods generally are more expensive per iterate as they often require more than a single function evaluation and the evaluation of one or more derivatives. Thus, although Halley's method has the lowest iteration count required for convergence, it has the highest cost per iteration since each iteration requires the evaluation of the function f and of its first and second derivatives. Thus one might ask students the question: is it computationally cheaper to use a higher order of convergence method, such as Halley's method, rather than a lower order method, such as the secant method to solve a problem? The students can investigate this question by generating a different type of image for each of the three higher order methods, with the iteration counts weighted by the cost per iteration relative to the secant method. The secant method is used as a basis for this comparison since it is the lowest cost iterative scheme we are considering (it requires only one function evaluation, one multiplication, one division, and three subtractions per iteration).

The iteration cost weighted plots for Newton's method, Halley's method and Aitken's method were generated by comparing the total cost to that of the secant method. The total cost is the startup cost plus the iteration count times the cost per iteration. For example, for the equation $x^7 - 1 = 0$, the startup cost for the secant method is 19 flops and the cost per iteration is 31 flops, whereas the startup cost of Halley's method is zero and the cost per iteration is 70 flops. So if for a particular starting iterate x_0 Halley's method required 8 iterations then the cost assigned to that x_0 would be 560 flops. This would correspond to between 17 and 18 iterations of the secant method and so we assigned color 18 to the pixel corresponding to x_0. Figure 2 is the result of this procedure applied to solving $x^7 - 1 = 0$. In this instance, one can see that all three methods (Newton, Halley, and Aitken) utilize the second gray scale, though the bands of colors have different widths. It is evident from this image that the total computational expense of using the higher order methods is about the same as for the secant method.

This is true in this case because simple polynomial functions have a low cost associated with evaluating the derivatives. However, if a more complex function is used, e.g., one that involves trigonometric, exponential or logarithmic functions, this would not necessarily be true. To illustrate this we considered solving the problem

$$f(x) = e^x + \sin(x) + x^2 = 0$$

in the region [-15,15] x [-15,15] with stopping criterion (1.2) and tolerance equal to 10^{-4}. It should be noted that this equation has an infinite number of roots since e^x has an infinite number of branches (or "copies") in the complex plane. Hence, the region of interest was chosen arbitrarily. Color Plate 4 is the composite image of the four fractals generated from our iterative methods by color coding the number of iterations needed for convergence. Once again we see that Halley's method has the lowest iteration count of the four as the majority of its image is colored with red hues. It should be noted that again Aitken's acceleration had a problem. The "bubbles" in the fractal are the result of floating point overflows in the norm calculations. Though this detracts from the beauty of the image, overflows are a common problem in iteration and must be dealt with.

When the cost per iteration consideration is included for this function (see Color Plate 5), it is clear that Halley's method is computationally more expensive than the secant method. Given that the blue regions correspond to high total computational costs while the red regions refer to low total computational costs, one can deduce from the colorings of Plate 5 that the secant method is the most cost efficient of the four techniques for this problem. Since the relative costs for different arithmetic operations are actually hardware dependent, the student can vary these values and look at the results.

The images presented thus far are useful not only in helping students understand the concepts of rate of convergence and cost of iteration but also the importance and effect of the stopping criterion on the iterative scheme. Though criterion (1.2) generates a more interesting image, it can lead to false solutions (as previously mentioned for the Aitken's acceleration method). Criterion (1.1), though more commonly used, can also be misleading. Once the function values of the iterates have become small we should be near a root of the equation. But which root? The same problem and resulting question arises when using almost any stopping criterion. Most students readily believe that the root to which the starting iterate x_0 is closest is the root to which the iterative scheme will converge. This is not always the case, especially if the starting iterate is approximately equidistant from two or more roots of the equation. To illustrate this situation we consider a different type of image representation of an iterative scheme. Rather than just color coding the number of iterations required for convergence we also code for the particular root to which the technique is converging. This is accomplished by assigning a color to each root of the equation and then using different shades of that color to indicate the number of iterations required for convergence.

An example of this is shown in Color Plate 6. This image was generated by applying Newton's method to the problem $x^6 - 1 = 0$ in the region [-2, 2] x [-2, 2] with stopping criterion (1.1) and tolerance equal to 10^{-4}. Once convergence was obtained (we started in a small enough region that this was guaranteed) the distance was computed from the final iterate to each of the six distinct roots of 1, which are

$$\{1.0,\ 0.5+(\sqrt{3}/2)i,\ -0.5+(\sqrt{3}/2)i,\ -1.0,\ -0.5-(\sqrt{3}/2)i,\ 0.5-(\sqrt{3}/2)i\}$$

where i is the square root of -1. Each of these roots was assigned a color: red, gray, green, light blue, purple, and dark blue (respectively). The root for which the distance from the final iterate was a minimum was chosen as the root to which the iterative process was converging and the corresponding color for that root was used to shade the pixel.

As above, the different shades of the various colors indicate the relative iteration count to convergence with darker shades indicating a lower iteration count. Examining this we notice the self-similarity structures along the rays containing the six roots of -1. If a starting iterate is located near one of these rays one can see from the colorization that the resulting iterates do not converge to one of the two closest roots and in fact the image becomes quite chaotic, as is more easily seen in Color Plate 8. The upper left image is the condensed version of Color Plate 6; the upper right is the representation of the region [$\sqrt{3}/2-0.2,\sqrt{3}/2+0.2$] x [0.3,0.7]; the lower left is the region [$\sqrt{3}/2-.02,\sqrt{3}/2$] x [0.485,0.505]; and the lower right is the region [$\sqrt{3}/2-0.005,\sqrt{3}/2-.002$] x [0.496,0.499]. As can be seen in these enlargements this chaotic nature remains when this portion of the image is magnified. This might be due to the use of the tangent line in Newton's method which could well "shoot" a starting iterate past the root closest to it resulting in the sequence of iterates converging to another root of the equation. This is speculation and more study needs to be done to determine why this phenomenon occurs. Recently, more research has been devoted to the characterization of the basins of attractors of fractals (for example [7] and the references made therein). Though this type of investigation can easily become highly complicated, studying this more closely might be a good student project — they may not find an answer but they could gain a deeper understanding of Newton's method.

Conclusions and Future Work

In this paper we have presented ways in which visualization may be used to aid the instruction of iterative techniques for solving the problem $f(x) = 0$. Fractal images for four methods are generated and presented simultaneously so that the important aspects of order of convergence and cost per iteration of these methods can be compared and more thoroughly understood. Also, an image colored by root was also presented to inform the student of the possibility of an iterative method converging to a possibly unwanted root. We have used such images in our numerical analysis and graphics classes and the student response has been excellent. This simple, yet elegant, use of visualization can easily be extended to other types of iteration, e.g., iterative techniques for solving linear and nonlinear systems of equations in two and three dimensions.

References

[1] Atkinson, Kendall E., *An Introduction to Numerical Analysis*, Wiley, New York, 1978.
[2] Pickover, Clifford A., "A Note on Chaos and Halley's Method," *Communications of the ACM*, 31(November 1988), 1326-1329.
[3] Stoer, Josef and Roland Bulirsch, *Introduction to Numerical Analysis*, Springer-Verlag, New York, 1980.
[4] Mandelbrot, Benoit, *The Fractal Geometry of Nature*, Freeman, San Francisco, 1984.
[5] Peitgen, H. O. and D. Saupe, eds, *The Science of Fractal Images*, Springer-Verlag, New York, 1988.

[6] Purdom, Paul W., Jr. and Cynthia A. Brown, *The Analysis of Algorithms*, Holt, Rinehart and Winston, New York, 1985.

[7] Alligood, Kathleen T. and James A. Yorke, "Fractal Basin Boundaries and Chaotic Attractors," in *Chaos and Fractals: The Mathematics Behind the Computer Graphics*, R. L. Devaney and L. Keen, eds., American Mathematical Society, Providence, RI, 1989.

Note on Software

The software to produce these images is written in Turbo C, for IBM PC compatible machines. The programs run in a noninteractive mode and generate an output file of iteration or cost values. This file is then read by another program which does the conversion from count/cost to pixel colors and displays the image. Thus, the display device can be different from the computational device. The images in the paper and the color plates are displayed on a Sun 386i with a color monitor but they can also be displayed on an IBM system with a VGA or a Mac II.

The computations are lengthy, e.g., each set of four images (computed at 750 x 750 pixel resolution) requires approximately ninety minutes to three hours on a 20 mhz 80386/80387 machine (running 16 bit software). By computing the images at a MAC II or IBM VGA compatible 480 x 480 pixel resolution this time drops to between twenty and forty minutes per image. By moving to a 32 bit compiler on an 80486 based machine the estimated execution time drops to around 5 minutes per image. Since this type of system will be common shortly, we are rewriting the software to make it interactive.

The student will be able to enter a complex function and choose from a set of numerical analysis techniques and quickly see the resulting images, both in iteration number and cost format. In addition, the student will be able to zoom in on a particular area of the plot by choosing a square area of the image (using an appropriate pointing device such as a mouse) and having the program recalculate this region. Thus, the student will be able to investigate particularly interesting, e.g., chaotic, areas of the plot. The current version of the software is available from the authors.

A Computer Graphics Approach to Stochastic Processes

David Griffeath

Random phenomena affect our science, technology, and daily lives in countless ways. From roots established in the seventeenth century, when Pascal and others developed the first systematic analysis of the laws of chance, probability theory has developed into a flourishing area of mathematical knowledge. Probability is the basic tool of statistics but also has applications in physics, biology, engineering, economics, and other applied sciences. Not surprisingly, then, courses in probability play an important role in the contemporary higher education curriculum.

The modern theory of probability is largely concerned with stochastic processes. *Stochastic* is just a fancy word for random, so stochastic processes model phenomena that evolve randomly in time. Think, for instance, of a small particle wandering randomly through space, a Geiger counter recording detections of some radioactive substance, or fluctuations in the length of a line of customers.

Whereas the interactive capabilities of computer software generally offer unique pedagogical advantages, the subject of stochastic processes is especially suited to graphics-based computer-aided instruction for at least three additional reasons. First, many of the models studied in probability are spatial and therefore more effectively represented by graphics than by literal description. Second, since the models evolve in time, they are most accurately pictured as random movies, that is, animations. Third, and most important, computers provide an efficient (fast) means of generating (pseudo-) randomness. Thus, computer graphics have the unique capability to display real-time simulations of the random dynamics being studied.

I have taught courses in probability and stochastic processes for more than ten years, and since the beginning have tried to incorporate visualization techniques into my instruction. At first this consisted primarily of crude chalk figures and ad hoc text-based simulations, in BASIC, of simple random experiments such as coin tossing. Later I began to use J. Laurie Snell's course and eventual book *Introduction to Probability* [4], which incorporates Monte Carlo simulations as an integral feature of the curriculum. But I was increasingly intrigued by the idea of a truly interactive color-graphics software library on stochastic processes.

So together with my graduate student Bob Fisch, I recently completed an educational software product entitled *Graphical Aids for Stochastic Processes (GASP)* [1]. The software runs on IBM PC-compatibles with color graphics. Our objective, which took about three years to fulfill, was to supplement traditional introductory courses in probability and stochastic processes with an extensive library of interactive computer graphic animations. The color graphics of *GASP* include so-called Monte Carlo simulations, more traditional

David Griffeath is Professor of Mathematics at the University of Wisconsin-Madison. He received his B.A. from Dartmouth College and Ph.D. from Cornell University. His research interests focus on the self-organizing dynamics of complex random systems and cellular automata. He actively incorporates interactive computer graphics experimentation in both his research and his teaching.

mathematical graphics, and several arcade-inspired educational games. *GASP* won the 1988 EDUCOM/ NCRIPTAL software competition in the Best Mathematics and Best Integrated categories.

The six modules of *GASP* treat the following models:

> Bernoulli trials
> Random walks
> The Poisson process
> Markov chains
> Branching & Queueing
> Brownian motion

The list is fairly standard; these models are the most basic building blocks of probability theory, those typically studied in a first course on stochastic processes. Some standard topics are gambler's ruin, Pólya's urn schemes, and the reflection principle. More contemporary topics are cellular automata, coupling, and random growth models. There are undoubtedly many additional topics in stochastic processes appropriate to graphics-based instruction.

I wish to make it quite clear that such visual aids are no substitute for the precise mathematical analysis developed in a traditional course in stochastic processes. On the other hand, my extensive teaching experience in this area has convinced me that effective visual representation of the models under investigation leads to more effective integration of probabilistic intuition and mathematical rigor. The interplay between intuition about the laws of chance and the formal axiomatics of probability theory is one of the most appealing aspects of the subject. Since many ideas concerning randomness are more effectively communicated through pictures than through words, it is my wholehearted belief that appropriate visual aids enhance the learning experience of today's students. The remainder of this article is intended to elaborate on my conviction. While many of my comments and examples will be based on experience with *GASP*, I hope they prove useful in a broader pedagogical context.

Instructional Issues

Randomness is a difficult concept. Weather reports, baseball statistics, and election polls give ample evidence of the use and abuse of probability theory in our daily lives. Everyone understands a little about chance phenomena, but no one understands the whole story. Students usually find the subject very challenging, even if they excel at other areas of mathematics. One needs to develop a keen intuition about randomness. Even though many aspects of probability, especially stochastic processes, have a substantial geometric component, in the traditional curriculum students are confronted with a barrage of set theoretic notation. A conscientious teacher will occasionally draw a rough sketch on the blackboard to explain the idea behind a probabilistic principle. But the student casualty list is high.

Where appropriate, computer graphics can offer students an alternative, visually-oriented perspective on the subject matter. I do not claim that this magically transforms stochastic processes into an easy subject. Rather, I am convinced that the extra intuition and broader context offered by visualization will benefit students in their inevitable confrontation with symbolism.

It is increasingly important for students of science and technology to understand basic principles of random dynamics. Stochastic processes software needs to be designed for a broad audience of students ranging from high school to graduate school, with interests in mathematics, statistics, engineering, and other applied sciences. Ideally, anyone should be able to learn a bit more about randomness. Clever high school students should be able to use it profitably, but the content should be sufficiently sophisticated that beginning graduate students also find plenty of challenging material. Mathematical ideas are not usually mastered at one particular level of formal education, but rather are assimilated through repeated exposure at several levels. There is even current impetus at the national level to introduce ideas about probability and statistics into the elementary school curriculum.

College probability courses are typically taught beginning at the upper undergraduate level. The target course for stochastic processes software will typically assume one semester of elementary probability as a prerequisite. Such service courses are often intended for students from many departments: mathematics, statistics, industrial engineering, operations research, business, economics, and others. The focus is on ideas and computation rather than the axiomatic development more appropriate to a graduate mathematics program. But graphics-based interactive software is appropriate to *all* levels of instruction in probability, statistics, and stochastic processes.

The Galton Board

Figure 1 shows the trajectory of a ball dropping through a regular lattice of pegs, a device sometimes called the Galton board after nineteenth-century English statistician Francis Galton. Such devices, frequently seen at science museums, effectively illustrate the ability of visualization to enhance our understanding of randomness. As each ball falls down from peg to peg, randomly bouncing off to the left or right, we can associate the ball's motion with a sequence of coin tosses by saying that a bounce to the right corresponds to heads and a bounce left to tails. When the balls accumulate at the bottom of the board they approximate a binomial distribution. On a big enough board the density profile predictably reproduces a Gaussian distribution, the famous bell-shaped curve.

Figure 1: *A ball's random path down the Galton Board*

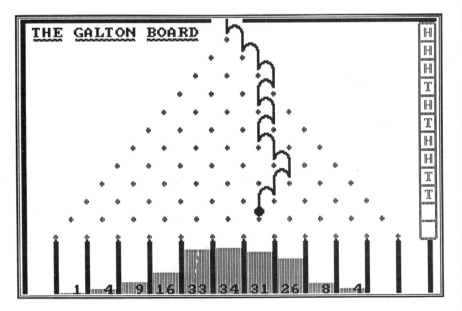

An animated Galton board provides an ideal visual introduction to probability and stochastic processes. By using a computer to produce random trajectories rapidly, and by allowing the user to interactively change the probability *p* that a ball falls to the right, we obtain a flexible environment for the simplest random experiments: so-called Bernoulli trials. This approach is used to study coin tossing, gambling, combinatorial identities, and the central limit theorem. Student response has been enthusiastic.

Let me describe in more detail a visual approach to combinatorial identities. With suitable indexing, the number of different paths down to the kth peg in the nth row of the Galton Board is simply the number of ways to choose a subset with k elements from a set of n elements. Thus, by labeling the pegs with these numbers, we obtain the celebrated Pascal's triangle. Virtually all of the important combinatorial identities that a student typically encounters in algebraic form can be interpreted geometrically in terms of this triangle. For instance, summing all the numbers in row n produces the middle entry in row $2n$. An interactive pattern recognition game effectively conveys these visual representations. Even if the goal is to master algebraic manipulation of binomial coefficients, the exercise of translating geometric patterns to their symbolic form is quite instructive. And the enterprising student will discover some patterns that are more naturally expressed by geometry than by algebra!

Finally, one can explore "paint by number" colorings of Pascal's triangle modulo n, for various choices of n, as a gentle introduction to the thoroughly contemporary topic of cellular automata. The case $n = 2$ produces Sierpinski's lattice, a discrete version of the fractal known as Sierpinski's gasket.

"Random" Numbers

Whereas the mathematical formulation of ideal randomness is well understood, methods for generating random outcomes on computers can only approximate the ideal. Random numbers supplied by many microcomputer programming languages are notoriously bad. A good visual test for uniform random numbers between 0 and 1 is to plot successive pairs (X, Y) as points on a graphics screen. Theoretically these points should gradually fill the screen with noise. The algorithm supplied with the original version of Microsoft BASIC produced plainly visible stripes when subjected to this test. The *random(n)* procedure of Turbo Pascal, to this day, gives horrendous results.

A colleague of mine recently discovered a bug in his favorite random number generator by visualization. He was simulating a random growth model for which one can prove that the occupied region spreads out according to a shape that is asymptotically a Euclidean ball. To his horror he discovered that some subtle error in the assembly language code produced an unsightly, persistent protuberance in the shape. This bug might well have eluded some of the standard statistical tests, but the message of the picture was clear.

Since the qualitative and quantitative accuracy of simulation depends critically on the accuracy of the underlying randomness, educational software in probability should use only carefully designed and tested algorithms. For *GASP*, we use two schemes. One is adapted from Knuth's authoritative account on pseudorandom number generation [2]. The other is based on a cellular automaton rule of Wolfram [5] that does well on the standard statistical tests. These rules appear to simulate accurately the stochastic processes that our software represents.

Software Design Considerations

There are a host of available programs that illustrate aspects of elementary probability and statistics, often incorporating conventional academic graphics. Line drawings, bar graphs, and the like are familiar ingredients in typical academic software. But state-of-the-art microcomputer graphics are found almost exclusively in computer games. I strongly advocate the use of what I like to call an arcade metaphor to render the basic models of stochastic processes in cartoon-like animations. Quality color graphics as shown in Color Plate 9 make a much more profound impression on the student. They also allow one to illustrate the modeling of random phenomena in ways not always directly related to the ostensible subject matter. For instance, in the queueing simulations of *GASP*, observant students will notice that the random people waiting for service execute subtle random walks as they cross the bank floor. In this manner, students are encouraged to reflect on the myriad ways that we experience randomness in our daily lives, and also to think about simple modeling of such random phenomena.

I have come to realize that the arcade metaphor is especially effective in animations that are actually presented as games. One example is the combinatorial pattern recognition game mentioned previously. Let me briefly describe three more.

- In a random quiz, students must decide whether sequences of *H*'s and *T*'s (heads and tails) are totally random or fake. The random sequences are generated by (pseudo-) ideal coin tossing. The fake sequences are generated by algorithms selected at random from a collection that deviate from the ideal in a variety of ways. For instance, some have too many heads, some have runs of heads and tails that are either too long or too short, and some have subtle symmetry properties. All the algorithms are capable of generating virtually any sequence, so students quickly learn that they cannot expect a perfect score. Nevertheless, scores improve noticeably over time as some of the subtleties of randomness are assimilated.

- A related game asks students to enter 100 random *H*'s and *T*'s. The computer secretly tries to predict each entry on the basis of the data entered up to that point. After each entry the student finds out whether the computer predicted correctly or not. By trying to outsmart the computer in this way, typical students are gradually able to reduce their scores from about 60-65 matches down to around 50, the expected score for ideal random entries. It is interesting to note that an optimal strategy based on knowledge of the computer's method could attain scores as low as 25 on occasion, but I have never seen a score below 40.

- In an attempt to convey something of the spirit of applied probability modeling, *GASP* also includes a simulation of a fast-food hamburger franchise. Students must hire the right number of workers for each shift, to maximize revenues in light of random customer arrivals that vary in intensity according to the time of day. Hiring too many servers is expensive, but if customers find that the lines are too long then they go next door for pizza. The business inevitably has good days and bad days. However one can dramatically increase average net profit by adopting the right schedule.

Users have consistently reported that such games are not only entertaining, but also among the most instructive aspects of the curriculum. Imagine the appetite for enjoyable graphics-based educational materials once the Nintendo generation reaches college age! Pedagogues horrified by the hyperactivity often associated with arcades should explore alternative, more contemplative approaches to interactive computer graphics, such as the Sierra On-Line recreational games [3].

I also advocate integrative software, combining the structured, sequential exposition of a tutorial with the flexibility and dynamic interaction of a simulation package. As a supplement to traditional courses in stochastic processes, such software can be used either in a computer lab or in classroom demonstrations aided by a video projection device. In either context, the interactive capabilities provide students with insights into the topic of randomness that are most effectively conveyed by computer graphics. This design allows the software to function either as a systematic tutorial to parallel a traditional course or as a collection of animated demos for occasional use in the lab or in class.

Laws of Averages

If asked to identify the most important theorem of probability and statistics, many experts will choose the central limit theorem, arguing that the key ingredient in randomness is variability. I would suggest, however, that laws of large numbers (commonly known as laws of averages) are even more fundamental: one cannot compute a variance without knowing the expectation first. The student's initial objective in any course about randomness is to understand what a probability is. The traditional curriculum tends to stress the frequency interpretation. (If you toss a coin long enough, then the proportion of heads will be close to 1/2.) Simulations can bring this perspective to life. Students can experience many examples of the law of averages in action along a single sample path. Once this connection between a physical incarnation of the model and its mathematical theory is made, computation-intensive analysis of fluctuations can be carried out by more conventional numerical techniques. Watching empirical frequencies converge to a theoretical limit carries two messages: on the one hand, the law of large numbers does really produce order out of chaos; on the other hand, a very large number of observations is required before this order becomes manifest.

A Detailed Example

Let me indicate the impact that visualization can have on students of probability by means of one specific example from *GASP*. A standard topic in any introductory probability course is *gambler's ruin*: starting with x, what is the probability of accumulating N before you go broke, by making $1 bets with probability p of success on each bet?

A typical text presents the problem in words, more or less as I did in the previous paragraph, and then proceeds to derive the theoretical solution using the symbolic calculations of theoretical mathematics. *GASP*, on the other hand, presents the problem quite concretely in terms of a toad jumping randomly along a highway, trying to reach a princess before it is incinerated by a dragon (see Figure 2). The user can vary the probability p of success (i.e., of jumping toward the princess) and see how this affects the toad's prospects. The user can also vary the starting position x and the goal N at will. Thus there are three parameters to the problem, and the user can interactively get a feeling for the dependence of the success probability on these parameters.

Some would say that this particular animation is merely a simulation. But the colorful toad, dragon, and princess convey a much more vivid sense of the problem than any descriptive prose or conventional academic simulation can offer. Now the instructor can derive the theoretical solution. But the role of visualization is not finished. A follow-up screen plots the theoretical success

Figure 2: *Toad's Ruin: an example of the arcade metaphor*

probability on a graph, shows a steady stream of hyperactive toads playing gambler's ruin, and keeps track of the (random) proportion that reach the princess. Students watch this proportion settling down to the mathematically predicted value (derived from finite difference equations) before their own eyes.

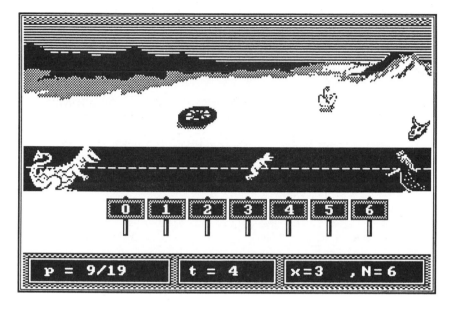

Visualization applications of this sort abound in stochastic processes. Graphics software can make a tangible connection between actual instances of randomness experienced on a computer screen and the rather elusive mathematical principles underlying them. Moreover, since students are selecting the parameters, the universality of these principles is confirmed in a manner unavailable through canned educational media such as textbooks, blackboard sketches, or even videotapes.

For in-class animations that take a long time, the instructor can introduce a lecture by starting a simulation, then blank the screen, lecture for an hour, and finally return to the animation at the end of class for the punch line. The element of suspense is very effective!

Impact on Students

Let me conclude this discussion with some representative student responses to *GASP*, collected as part of course evaluations. To me, they indicate the great potential of visualization in stochastic processes education, and in mathematics instruction as a whole:

... It is simply a lot easier to look at formulas, read a textbook, and understand what's going on if I can attach a picture to it. I can remember the material substantially longer if I can attach a picture to it. ...

... If I had to learn about this subject wholly through use of the textbook I would be in big trouble. I feel that the software makes the class much more interesting. I think that the subjects are fairly complex, but the software makes the class very manageable. ...

... It makes me understand better, especially when I can change the variable. Then I can see how each variable has effects on the process. ...

... GASP has contributed greatly to my understanding of the theory of stochastic processes by making the concepts easy to grasp. In most cases the time spent on GASP has been worthwhile; however I think more traditional materials would be more effective for learning the notation involved. ...

... GASP brings some of the more esoteric theorems and concepts back to a level where one can get some idea of how they work in practice, something many textbooks do not do. ... I like the fact that it is easy to use, even for a computer klutz like myself. ...

... There is a saying, 'I see, then I believe.' So I believe, then I enjoy. ...

References

[1] Fisch, R. and D. Griffeath. *Graphical Aids for Stochastic Processes. An Interactive Tutorial in Six Modules.* Wadsworth and Brooks/Cole Advanced Books & Software, Pacific Grove CA, 1989.

[2] Knuth, D. E. *The Art of Computer Programming. Volume 2: Seminumerical Algorithms.* Addison Wesley, Reading, MA, 1969.

[3] Sierra On-Line, Inc. P.O. Box 485. Coarsegold, CA, 93614.

[4] Snell, J. L. *Introduction to Probability.* Random House/Birkhauser, New York, 1988.

[5] Wolfram, S. "Random sequence generation by cellular automata." *Adv. Appl. Math.*, 7(1986), 123-169.

Dynamic Visual Experiments with Random Phenomena

Julian Weissglass and Deborah Cummings

There is increasing evidence that even successful students in traditional mathematics programs do not have a deep understanding of mathematical concepts. This is worrisome to many educators. *Everybody Counts* [5] is one of the most recent documents concerned about mathematics education in the U.S. In discussing this problem it states: "Students simply do not retain for long what they learn by imitation from lectures, worksheets, or routine homework. Presentation and repetition help students do well on standardized tests and lower-order skills, but they are generally ineffective as teaching strategies for long-term learning, for higher-order thinking, and for versatile problem-solving."

Courses in statistics and probability are not exempt from this criticism. In fact, the research of Pollatsek, et. al., [6] shows that students complete introductory statistics with little ability to apply it to either future courses or the real world. They state: "One pedagogical point seems clear: in many introductory courses students are taught to use formulas in a rote manner with the justification that thorough understanding of the material can wait until the second course or later. While it is undeniably true that students can solve some problems with this approach, our data suggest that the range of problems that can be solved with only instrumental knowledge is vanishingly small."

Shaughnessy writes [8] that "Some of our misconceptions of probability may occur just because we haven't studied much probability. However, there is considerable recent evidence to suggest that some misconceptions of probability are of a psychological sort. Mere exposure to the theoretical laws of probability may not be sufficient to overcome misconceptions of probability."

For mathematics educators possible remedies to the lack of understanding come readily to mind. Pollatsek, et.al., suggests that pictorial representations may be more useful in helping students' learning than verbal or written descriptions. Shaughnessy recommends that, "Introductory probability and statistics be activity-based experimental probability" and "Emphasis should be placed on simulation, both as a tool to model experiments and as a problem solving technique." Using the computer to perform experiments and provide a pictorial representation of the results is an obvious linkage of these two recommendations. Although the use of computers in statistics courses has grown rapidly, the predominant use has been as a computational (using statistical packages) rather than an instructional tool.

This paper will describe one approach to using visualization in learning statistics. The approach uses the ability of computers to perform statistical experiments (with the parameters determined by the learner) and display the

Julian Weissglass is Associate Professor of Mathematics at the University of California, Santa Barbara. He has experimented with alternatives to the lecture method of teaching mathematics at the college level for the past 20 years. His article, "Small Groups: An Alternative to the Lecture Method" won the George Pólya award from the MAA in 1977.

Deborah Cummings is a secondary mathematics teacher and mathematics coordinator in Fillmore Unified School District. She has worked extensively in professional development at the secondary level in mathematics.

results dynamically (as they occur). We will refer to the method as *dynamic visual experimentation*. We will use the term random phenomena to refer to the broad class of situations whose mathematical analysis requires statistical or probabilistic concepts or methods.

We will explain
- how one set of computer programs has implemented dynamic visual experimentation with random phenomena,
- the epistemological basis for the approach,
- the pedagogical and curricular implications of using dynamic visual experiments in statistics, and
- the results of a small controlled experiment using the software.

Dynamic Visualization Programs

Fifteen computer programs (for the Apple II and IBM PC compatibles) were developed as part of a courseware package entitled *Hands-On Statistics* [12]. Eleven of the programs allow the user to do sampling (with replacement) experiments and display the results graphically. A tabular display of the results can also be selected. The programs allow the user to choose the appropriate parameters, such as sample size, number of trials, the "probability of a success" or the population mean. The instructor can easily construct normally distributed populations with which students can experiment. In addition there are four programs that allow the user to experiment graphically with statistical concepts (histograms, means, standard deviation, regression lines, correlation coefficient, z-scores, normal curves, etc.) but do not involve random sampling.

It is somewhat of a contradiction to explain dynamic phenomena in the static medium of the printed page. An educator can gain a true appreciation of both the visual effects and the interactiveness of these programs only by using them. We will, however, explain a few of the programs in an attempt to provide a flavor of what they can do.

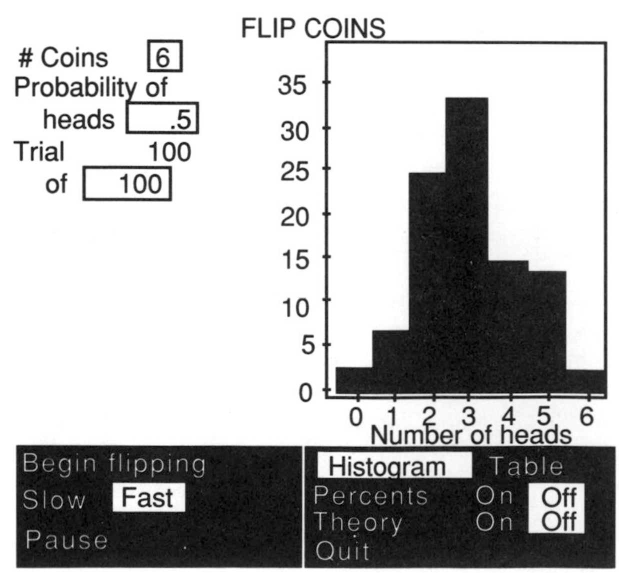

FLIP COINS allows repeated experiments of coin flipping and displays the histogram after each trial. When flipping a single coin, the results of the first 100 flips are shown on the screen. The user can change the number of coins flipped as well as the probability of getting a head. Figure 1 shows the results of flipping 6 fair coins 100 times. (Figures have been reproduced for this article.)

The theoretical distribution can be superimposed on the histogram, thus allowing the user to see that the larger the sample, the closer the actual distribution gets to the theoretical. The program BINOMIAL extends FLIP COINS by allowing the student to compute the area of the histogram to the left of a movable vertical line. BINOMIAL will also superimpose the approximating normal curve. By choosing larger and larger sample sizes one can see how the normal approximation to the binomial distribution gets better and better.

DRAW TICKET simulates a lottery. It allows the student to label from 1 to 30 tickets and simulate drawing with

■ **Figure 1**

■ **Figure 2**

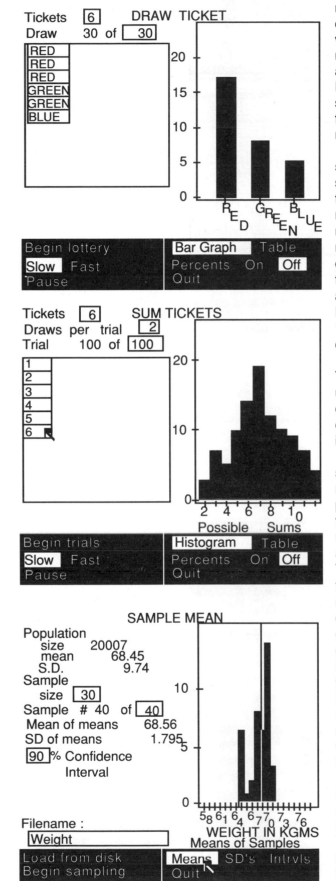

■ **Figure 3**

■ **Figure 4**

replacement. It provides experience with elementary notions of probability — the variability of results and that the actual results tend to be closer to the theoretical as the sample size becomes large. Figure 2 shows the results of DRAW TICKET simulating 30 draws with replacement from a jar with 3 red marbles, 2 green marbles, and 1 blue marble.

SUM TICKET is a more sophisticated lottery simulation program. It allows the student to simulate random processes that can be modeled by drawing (with replacement) numbered tickets and summing the results. (See [3] or [12] for an explanation.) The student can specify up to 40 tickets, the number of draws per trial, and the number of trials. For example, in Figure 3 there are six tickets labeled 1,2,3,4,5,6 with two draws per trial so the experiment simulates rolling two dice.

The workbook for *Hands-On Statistics* uses SUM TICKET to explore the concept of expected value through simulating such experiments as roulette wheels and investigating de Méré's paradox.

SAMPLE MEANS allows the student to take repeated samples of a given size from a population which is loaded from a file on the disk. The program computes the mean and the standard deviation as well as a confidence interval for each sample. The student can select what to display on the screen — either the histogram of sample means, the histogram of standard deviations, or the confidence intervals corresponding to each sample. The set of confidence intervals (one for each sample) is drawn on a graph with a line indicating the population mean. For example, Figure 4 shows the histogram of sample means obtained by taking 40 samples of size 30 from a population of 20,000 weights. Figure 5 shows the confidence interval for each sample.

In SAMPLE MEANS the user can determine the size of the samples as well as the number of samples taken and the confidence level. The student can see how sample size affects the length of the confidence interval and the distribution of means.

■ **Figure 5**

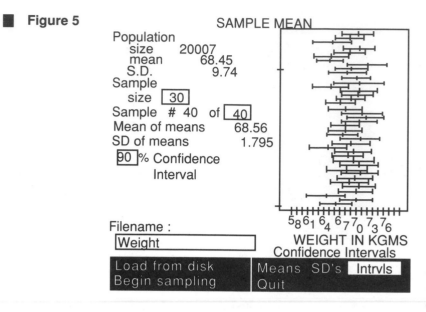

Furthermore, he or she can verify that the confidence level does give an approximation to the percentage of samples yielding confidence intervals that capture the population mean.

TEST TWO MEANS allows the student to sample from two populations (or the same population twice). The student selects the sample size after loading the populations from the disk. The computer does the sampling from each population in turn and displays both histograms on the screen. If the same population is loaded twice, this is an excellent way of comparing how sample size affects the distribution of sample means. The program also computes and displays the difference of the two sample means and computes the probability of obtaining an absolute difference of the two means as large as the one obtained given the hypothesis that the two means are equal. Figure 6 shows the results of testing the hypothesis that the mean price of a house in San Diego and Seattle are the same.

■ **Figure 6**

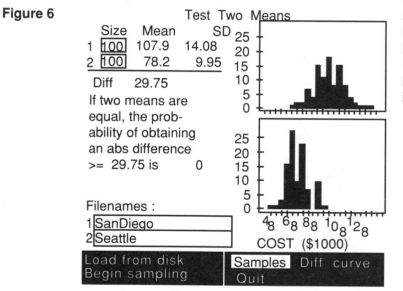

Learning and Dynamic Visual Experimentation

To a large extent the current mathematical reform effort at the precollege level has been influenced by the epistemological school called constructivism. "From the constructivist perspective, learning is the product of self-organization" [9]. *Everybody Counts* [5] supports this position, stating: "Educational research offers compelling evidence that students learn mathematics well only when they *construct* their own mathematical understanding." The process of constructing understanding requires the active engagement of the learner in making sense of a situation.

We will make some conjectures about how dynamic visual experimentation might help students construct their understanding of random phenomena. Although we will report on some positive results for students using dynamic visual experiments later on in this paper, considerable additional research would be required to verify these conjectures.

Conjecture 1. *Performing experiments, while in control of parameters, is beneficial to learning.* Having the power to set the parameters requires the student to be engaged, at least to some extent, and stimulates interest. Students enjoy watching what happens on the computer screen when they change the parameters of an experiment or even do the same experiment again. There is an inherent human curiosity about the effects of one's actions ("What if I do this?") and the software can tap into this curiosity to assist learning. We conjecture that more passive ways of seeing graphical representations of random phenomena (for example, watching a sequence of preprogrammed computer experiments or watching a movie or a video) would not have the same effect.

Conjecture 2. *Visual representations of concepts and processes assist understanding.* This is the reason that textbooks have graphics and teachers encourage math students to "draw a diagram." The computer can do the calculations and draw the pictures faster than humans can. In fact, the computer can display simulations dynamically — as they occur.

Conjecture 3. *Being able to repeat experiments at will is beneficial.* Repeatedly seeing the histogram of the sample evolve will enable students to grasp intuitively the notion that samples, although they vary, can nevertheless be used to estimate the statistics of a population.

Conjecture 4. *The possibility of visually comparing the results of experiments with large sample size to experiments with small sample size is beneficial in understanding how estimation improves as the sample size becomes larger.* Several of the programs in *Hands-on Statistics* allow sampling from a population. Changing the size of the sample and comparing the distributions should lead to increased understanding of how sample size affects the accuracy of any prediction. Furthermore, changing the parameters for a binomially distributed population will enable students actually to see that, for all values of p, if the sample size is large enough the normal curve is a good approximation.

Conjecture 5. *It is the combination of the capacity to experiment and to see the graphic representation of the results immediately on the screen that enhances understanding.* It is fairly simple to write programs in BASIC or LOGO that simulate sampling. It is also possible to view films or videos of simulations depicted graphically. The unique feature of dynamic visual experimentation is that the student can control the experiment *and* see the results graphically.

Implications for Pedagogy and Curriculum

Although we believe that method and content are inextricably bound, we will address them somewhat separately in what follows.

It is clear that the lecture method is the prevalent method of instruction in college classrooms. It continues to be so despite the fact that research "points to the conclusion that current mathematics education does not adequately engage students' interpretive and meaning-construction ability" [7]. The lecture method prevails even though articles describing the use of small group methods at the college level have been available since 1971 (see [1], [2], [10], [11]). With the lecture method, most college students only become actively engaged in doing mathematics when they do their homework. That activity, however, especially in introductory courses, often consists of students figuring out which method presented in class can be applied to a particular problem. There is little opportunity to discuss mathematics, make decisions about what to try next, or state an opinion about a mathematical situation. An instructor may say that, "you only learn mathematics by doing mathematics," but students will not come to appreciate the nature of mathematics from listening to lectures and doing homework. As Marshall McLuhan emphasized, "The medium is the message" [4] and this is as true in education as anywhere else. Sitting passively listening to instructors talk about mathematics gives most students the message that mathematics is a collection of rules and theories devised by others that they need to memorize.

Dynamic visual experiments do not by themselves magically provide all the instructional strategies that we would like to see in a mathematics classroom. By actively involving students in making and testing conjectures and interpreting the results of experiments, however, they can provide a basis for a more constructivist approach in the classroom. Such an approach will engage students in the processes of mathematics (conjecturing, generalizing, analyzing, deducing, interpreting, for example) and in communicating with each other about mathematics. Instructors can drastically reduce the amount of time they spend lecturing to students. Curriculum can be devised that allows students to work in small groups in a computer lab. They can discuss situations that involve random phenomena and use the computer to do experiments to test their hypotheses. Although many instructors are using graphical displays as demonstrations in lecture classes, this does not meet the need for students to do and interpret experiments for themselves.

We do not think that dynamic visual experiments replace the need for hands-on physical experimentation, even at the college level. Physical experiments are limited, however, by the inconvenience of bringing materials to class, the time required to take large samples, and the tedium of organizing the data. The computer provides a tool for taking large sample sizes and visually displaying the data.

Another way of considering the implication of dynamic visual experiments for the classroom is related to the issue of play. The software provides the opportunity for "playing" in the sense that play is characterized by the learner being in charge and having fun. The ease with which the learner can use the software to ask "what if?" questions is an important factor in providing a playful atmosphere for learning. Being able to see the results, however, is critical. Visualization contributes greatly to the playfulness inherent in using the software.

The availability of any type of new technology raises the question: "How should the content of traditional courses be changed to take account of the new tools?" This is a familiar question for mathematics educators. Some questions have been settled: we don't teach the square root algorithm any more, since calculators with square root keys are available. Some questions are still disputed: some high school algebra teachers insist that trigonometry students use tables rather than calculators when working with trigonometric functions. Some questions are just being posed: do we continue to spend weeks in first year calculus developing student skills in graphing functions when graphic calculators and computers can do the job so much more easily?

Instructors in probability and statistics courses need to think about how the availability of the computer for dynamic visual experiments affects the course content. In many courses the goal is not a rigorous understanding of the mathematical theory of probability and statistics. In these courses graphical programs can provide the necessary understanding for students to be able to use statistical analysis packages in meaningful ways. Even in courses with mathematical rigor as a goal, dynamic visual experiments could enhance understanding. In an introductory course, one must question whether it makes any sense at all to present the formulas for mean, standard deviation, correlation coefficient, confidence intervals, etc.

Another implication for curriculum is that methods for dealing with random phenomena are accessible to students at an earlier age than previously. High school students can use these graphical programs to perform experiments and think about the mathematical implications without having to memorize formulas or grapple with notation.

Experimental Results

In order to discover whether using these programs would enhance understanding of statistical and probabilistic concepts, one author (Cummings) conducted a study with high school students. She taught a four week unit on probability to three Algebra II classes. All students received similar instruction. One class, the experimental class, worked in in a computer lab using three programs (FLIP COIN, DRAW TICKETS, and SUM TICKETS) from *Hands-On Statistics*. The students from the experimental class worked in pairs during class time. The instructional time was the same for all students. All classes were pre- and posttested with the eight questions shown in Figure 7. Shaughnessy used questions 2 through 6 in his study [8] of misconceptions about probability.

The two classes that did not use the computer programs were grouped as the control group. The overall mean scores and standard deviation on the pre and post test were as follows:

	Pretest Mean	Pretest SD	Posttest Mean	Posttest SD
Control group (n=38)	2.29	1.29	2.47	0.89
Experimental group (n=17)	2.94	1.14	3.82	0.88

The experimental group's mean gain was .88 as compared to the control group's mean gain of .18, and both the gain of the experimental group and the difference in the gains between the two groups are statistically significant at the 5% level. One-tailed t-tests were conducted for the differences in pre- and posttest means within each group. The change in mean for the control group was not significant ($p = .158$). The change in mean for the experimental group was significant at the 5% level ($p = .015$). Because the sample size of the experimental group was small a nonparametric test (the Wilcoxon matched-pairs signed-rank test) was also conducted. The change in mean was again found to be significant at the 5% level ($p = .049$). We also tested whether the changes in scores of the two groups were different using a nonparametric test. The change in student test scores were computed. The null hypothesis was that these two groups came from the same population. This hypothesis was tested with the Mann-Whitney U Test and rejected at the 5% level ($p = .049$).

Figure 7: *A test on statistical and probability concepts*

1) You have tossed a "fair" coin 20 times and gotten heads every time. What is most likely to happen on your next toss?
   a) heads   b) tails   c) both equally likely.

2) The probability of a baby's being a boy is about 1/2. Which of the following sequences is more likely to occur in having six children?
   a) BGGBGB   b) BBBBGB   c) Both about equally likely.

3) What is the probability that among 6 children 3 will be girls?

4) The probability a baby will be a boy is about 1/2. Over the course of an entire year, would there be more days when at least 60% of the babies born were boys.
   a) in a large hospital?   b) in a small hospital?
   c) it makes no difference?

5) Consider the grids below.

   A.   X X X X X X X X X   B.   X X
          X X X X X X X X X   X X
          X X X X X X X X X   X X
                                              X X
                                              X X
                                              X X
                                              X X
                                              X X
                                              X X

   Are there:
   a) more paths possible in grid A?
   b) more paths possible in grid B?
   c) about the same number of paths in each?

6) A person must select committees from a group of 10 people. (A person may serve on more than 1 committee.) Would there be
   a) more different possible committees of 8 people?
   b) more different possible committees of 2 people?
   c) about the same number of committees of each?

7) A die is rolled. What is the probability that the die will show an even number?

8) A fair coin is tossed 3 times. What is the probability of at least 1 head?

We certainly do not know why the experimental group did better on the test. It could have been the opportunity to visualize statistical phenomena, increased student motivation from using computers, having the opportunity to perform experiments, or a sense of being in charge of one's learning. Most likely it was some combination of these. A further study would be necessary to attempt to isolate the various causes and to investigate the conjectures made earlier. The assessment instrument used was not specially designed to assess the effects of visualization on learning. A more sophisticated instrument might be able to determine whether any of the conjectures made above are valid.

Conclusion

If understanding random phenomena is the goal, having students memorize formulas and how to use them to obtain numerical results may not be the best path to follow. Many mathematics educators recommend that students participate in hands-on experiments in order to develop their understanding.

We agree with this recommendation but would go further. We recommend that as an accompaniment to any theoretical investigation, students have the opportunity to perform experiments with a computer where the results of these experiments are displayed graphically and where the student is in charge of the parameters of the experiment.

References

[1] Davidson, N., "The Small Group Discovery Method as Applied in Calculus Instruction," *Amer. Math. Monthly*, (1971), 789-791.

[2] Davidson, N., "The Small Group Discovery Method in Secondary and College Level Mathematics," in *Cooperative Learning in Mathematics, A Handbook for Teachers,* N. Davidson, ed., Addison-Wesley, Menlo Park, CA, 1989, 335-361.

[3] Freedman, D., R. Pisani, and R. Purvis, *Statistics*, W.W. Norton and Company, New York, 1974.

[4] McLuhan, M., *Understanding Media: The Extensions of Man*, McGraw-Hill, New York, 1964.

[5] National Research Council, *Everybody Counts: A Report to the Nation About the Future of Mathematics Education*, National Academy Press, Washington, D.C., 1989.

[6] Pollatsek, A., S. Lima, and A. D. Weil, "Concept or Computation: Students' Understanding of the Mean," *Educational Studies in Mathematics*, 12(1989), 191-204.

[7] Resnick, L. B., *Education and Learning to Think*, National Academy Press, Washington, D.C., 1987.

[8] Shaugnessy, J. M., "Misconceptions of Probability: From Systematic Errors to Systematic Experiments and Decisions," in *Teaching Statistics and Probabilty,* A.P. Shulte, ed., National Council of Teachers of Mathematics, Reston, VA, 1981.

[9] von Glasersfeld, E., "Cognition, Construction of Knowledge, and Teaching," *Synthese*, 80(1989), 121-40.

[10] Weissglass, J., "Cooperative Learning Using a Small Group Laboratory Approach," in *Cooperative Learning in Mathematics, A Handbook for Teachers,* N. Davidson, ed., Addison-Wesley, Menlo Park, CA, 1989, 295-333.

[11] Weissglass, J., "Small Groups: An Alternative to the Lecture Method," *Two Year College Math. J.*, 7(February 1976), 15-20.

[12] Weissglass, J., N. Thies, and W. Finzer, *Hands-On Statistics: Explorations with a Microcomputer*, Wadsworth, 1986.

This volume was set with Aldus Pagemaker 3.02 on the Apple
Macintosh, based on designs by B. Longo Associates in Turlock,
California. The body of the text is in the Helvetica font with
Helvetica Black and Helvetica Black Italic titles and headers.
Greek and mathematical characters are in the Symbol font, and
many of the formulas were set in MathType 2.0 with the Belmont
font.